REEF LIFE

Front Cover
Zebra Crab (*Zebrida adamsii*) with
venomous long-spined sea urchin (*Astropyga radiata*)
Back Cover
Top left: Pygmy Seahorse (*Hippocampus bargibanti*) with gorgonian (Melithaeidae)
Top right: Pink Skunk Anemonefish (*Amphiprion perideraion*) with
Magnificent Sea Anemone (*Heteractis magnifica*)
Bottom left: Blue-green Chromis (*Chromis viridis*) with
staghorn coral (*Acropora* sp.)
Bottom right: Barredfin Moray Eel (*Gymnothorax zonipectus*) in sponge with
Common Cleaner Shrimp (*Lysmata amboinensis*)

Produced and distributed by
T.F.H. Publications, Inc.
One T.F.H. Plaza
Third and Union Avenues
Neptune City, NJ 07753
www.tfh.com

REEF LIFE

NATURAL HISTORY AND BEHAVIORS OF MARINE FISHES AND INVERTEBRATES

TEXT BY
Denise Nielsen Tackett

PHOTOGRAPHS BY
Denise Nielsen Tackett & Larry Tackett

FOREWORD BY
Ronald L. Shimek, Ph.D.

T.F.H. Publications
One T.F.H. Plaza
Third and Union Avenues
Neptune City, NJ 07753
www.tfh.com

ISBN 1-890087-55-6 (hardcover); ISBN 1-890087-56-4 (softcover)

Printed and bound in the United States of America

Library of Congress Cataloging-in-Publication Data

Tackett, Denise Nielsen.
Reef life: natural history and behaviors of marine fishes and invertebrates / text by Denise Nielsen Tackett; photographs by Denise Nielsen Tackett & Larry Tackett; foreword by Ronald L. Shimek.
p. cm.
Includes bibliographical references (p.).
ISBN 1-890087-55-6 (hc)—ISBN 1-890087-56-4 (sc)
1. Coral reef animals. I. Tackett, Larry. II. Title.
QL125.T33 2002
591.77'89—dc21 2001059063

Color separations by T.F.H. Publications, Inc.
Designed by Eugenie Seidenberg Delaney

Co-published by
Microcosm Ltd.
P.O. Box 550
Charlotte, VT 05445
www.microcosm-books.com

TO OUR PARENTS

Norma & Denis *Edith & Oliver*

for instilling in us the values
necessary to recognize that
which is truly important in life.

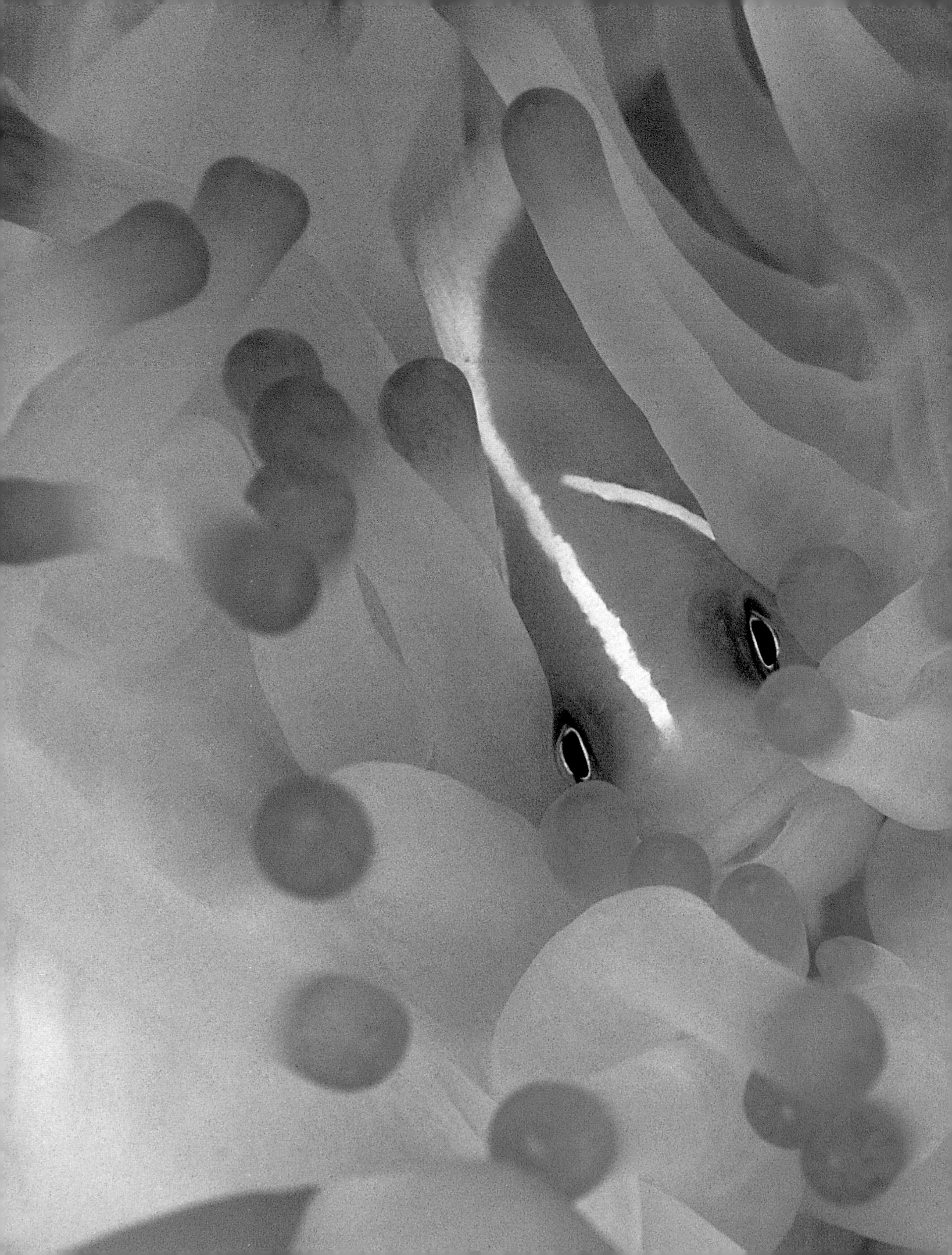

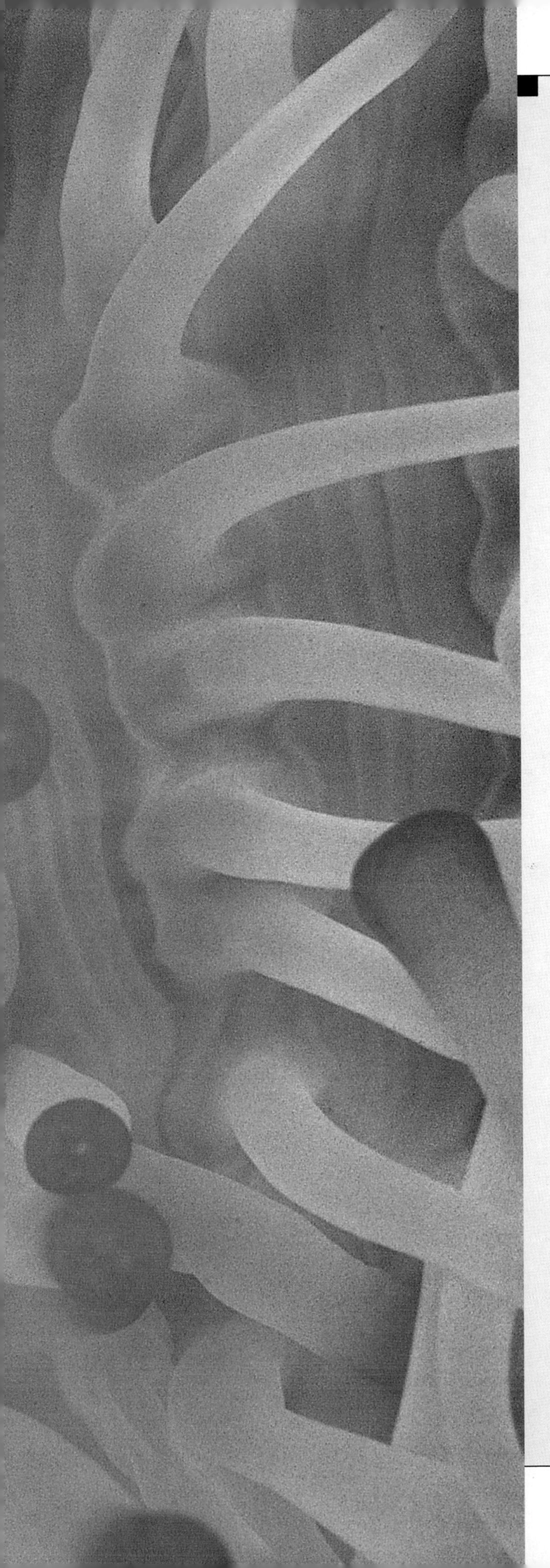

CONTENTS

Symbionts: Pink Skunk Anemonefish (*Amphiprion perideraion*) seeks lifelong shelter in its host sea anemone (*Heteractis magnifica*).

ACKNOWLEDGMENTS

WE WISH TO THANK ROGER STEENE, FRED Bavendam, and especially Scott Michael for their encouragement and support, without which this book would never have been written.

Dave Behrens, Larry Jackson, and Ronald L. Shimek were all kind enough to review the manuscript and offer their suggestions, for which we are deeply grateful. We also thank Susan Ritman Macdonald for her invaluable comments.

We are indebted to a number of recognized world experts, researchers, and scientists who have been kind enough to share the fruits of their labors with us over the years. These include Jack Randall, Amanda Vincent, Mark Norman, Sara Lourie, Roger Hanlon, Eric Hochberg, Rudie Kuiter, Harry Erhardt, Charles Anderson, Chou Loke Ming, Carden Wallace, Bill Rudman, and Mark Erdmann.

We gratefully acknowledge and appreciate the hospitality and support extended to us by: Kathryn Ecenbarger, Mark Ecenbarger, and the staff of Kungkungan Bay Resort, especially the sharp-eyed dive guides, Lembeh Strait, Indonesia; Marlene Frost, Graham Frost, Adam Frost, and Southeast Asia Divers, Phuket, Thailand; Lorenz and Renée Mäder and Wakatobi Resort, Sulawesi, Indonesia; Matthew Hedrick and Dive Asia Pacific, Phuket, Thailand; Ismail Hilmey, Halaveli Resort and the Coral Princess, Maldives; Giorgio Rosi Bellière and Seafari Adventures, Italy and Maldives; Max Ammer and Irian Diving, Sorong, Indonesia; Daniel Johannes Gondowidjojo, Tasik Ria Resort, and Eco Divers, Manado, Indonesia; Emiko Shibuya and Dive Paradise Tulamben, Bali, Indonesia.

We'd like to acknowledge the additional assistance provided to us in our travels by Dan Marino of Singapore Airlines, Silk Air, Nancy Gimblin, Nancy Nevis, Ken Knezick, and Dominick Macan.

On a personal note, we wish to acknowledge some people who provided us with encouragement and moral support in good times and bad: Larry Smith, our old diving buddy, and our friends Ursula Kara, Greg Gapp, and Jeremy Barnes.

We'd also like to recognize the Arizona State University Cancer Research Institute, especially G.R. Pettit. We did part-time field work for the Institute for 13 years. During that time we had the good fortune to work and dive in 11 different countries, which gave us an extraordinary opportunity to observe marine animals over a wide geographic area.

A special thanks goes to our sponsors: Toshi Kozawa and Junko Maruoka of Anthis Underwater Photographic Equipment; Sandy Thomas and Ikelite Underwater Systems; and Dave Reid and Terry Schuller of Ultralight Control Systems. Thanks also to James Lyon and Kodak for their support.

We thank our publisher, James Lawrence, from the bottom of our hearts for taking a chance on us and for giving us the freedom to realize our vision. We are indebted to Eugenie Seidenberg Delaney for her handsome design, and Alice Lawrence and Alesia Depot at Microcosm, Ltd. for their professionalism and attention to detail.

Finally, and most of all, we are grateful to Mother Nature for giving us such a wonderful environment in which to work. ■

Shoaling: Oriental Sweetlips (*Plectorhinchus orientalis*) cluster for mutual protection and group foraging.

Foreword

LOOK CLOSER

Ronald L. Shimek, Ph.D.

"In the biting honesty of salt, the sea makes her secrets known to those who care to listen."

SANDRA BENITEZ,
A PLACE WHERE THE SEA REMEMBERS (1993)

ON EVERY REEF TRIP THERE ARE AT LEAST TWO and very occasionally three distinct kinds of divers and snorkelers. First, most of us have encountered those whose primary goal underwater is to go from Point A to Point B—as fast as possible. As a marine zoologist, I have coined a scientific classification for these folks: "scooby divers." Their sole purpose in diving seems to be the thrill of swimming underwater. When they do occasionally take note of an animal and recognize it as such, they will make the statement that goes something like, "Oh yeah, I saw a fish."

Another type of underwater visitor, probably constituting the majority of all scuba divers and snorkelers, is genuinely interested in looking at all sorts of things underwater. However, this diver or snorkeler has, at best, a rudimentary understanding of the natural history and biology of a reef and has never been taught the basic observational skills needed to appreciate the world that he or she is swimming in.

The final type of diver is really quite rare, and I put myself and the authors of this book in that category.

Reef in miniature: Red Soldierfish (*Myripristis vittata*) and Ribbon Sweetlips (*Plectorhinchus polytaenia*) shelter under a large table coral (*Acropora* sp.) in a small section of reefscape that reflects the complexity of one of Earth's most dazzling ecosystems.

1 Bluestreak Fusilier (*Pterocaesio tile*)
2 Black and White Snapper (*Macolor macularis*)
3 Reticulated Dascyllus (*Dascyllus reticulatus*)
4 Table coral (*Acropora* sp.)
5 Ribbon Sweetlips (*Plectorhinchus polytaenia*)
6 Sponge (Porifera)
7 Cup corals (*Tubastraea* sp.)
8 Red Soldierfish (*Myripristis vittata*)
9 Soft coral or sponge with zooids
10 Plate coral (*Montipora* sp.)
11 Tailspot Squirrelfish (*Sargocentron caudimaculatum*)

This last class of reef explorers encompasses those rare human fish who have the patience and the desire to see—really see—their underwater surroundings. These are the folks who can spend their entire dive exploring a single coral outcropping for the minute details and the unique plants and animals that actually create the wonderful world of a coral reef. The underlying mission of this extraordinary book, as I see it, is to coax a new generation of amateur naturalists into the classification of true students and explorers of the reef.

Coral reefs are among the most biologically diverse habitats on our planet, and they harbor examples of virtually all types of organisms, with the blessed exception of insects. Quite a few of the myriad species found on reefs are large and obvious, but many, many more are small, obscure, or reclusive.

Close and careful observation of smaller things in underwater marine habitats will often tell you a great deal about the habitat that isn't obvious from a distance. Human beings are large animals, and we tend to overlook the small and the seemingly insignificant, but the struggle of life on a coral reef is waged on the scale of things large and small, and is often more frequently seen and more intense in the smaller size ranges.

PREDATOR WATCH

Looking closely at the plants and animals of a small area can provide many surprises and telltale signs about the life of the larger system. Predation is one of the most intense biological interactions, and yet actual observations by humans of natural predation in many ecosystems are very rare. Ask yourself, how many times in nature have you seen an animal kill and eat another animal? Unless you have been diving or snorkeling, such observations are often vanishingly few. However, this is not the case in most underwater habitats. Here, you just have to slow down and look closely to see predatory events all around you.

Some years ago, I was doing research in an area dominated by hydroids. To a casual observer, these animals appear to be a reasonably permanent feature of the substrate; they are colonial, rather plantlike in appearance, and seemingly untouchable because of their known stinging abilities. From a distance of several feet away, these reef fixtures might not merit a second look. However, I found during one close inspection that there were more than 5,000 predatory nudibranchs per square meter eating the hydroids. Each nudibranch was a few millimeters long, and from a short distance away, they were unnoticeable. Nonetheless, this sort of intensive predatory pressure is an underpinning of the whole ecological community.

But you don't have to be out on a reef to appreciate the value of being a better observer. Like divers, many of my fellow marine aquarists are equally guilty of never slowing down and truly seeing the captive worlds sitting before their eyes. Just as many divers are intently focused on spotting a barracuda or other large pelagic species, many aquarium keepers have eyes only for bigger fishes and centerpiece corals.

However, some of the most fascinating scenes ever played out in a marine aquarium take place in the

More than meets the eye: Schooling Bannerfish (*Heniochus diphreutes*) form a typical shoal or same-species aggregation, sweeping over the reef to feed on plankton. These fish exhibit both survival behavior (schooling) and genetic adaptation (disruptive coloration).

niches of minireefs that so many hobbyists are now maintaining in their home systems. A perfect example is the Peppermint Shrimp (*Lysmata wurdemanni*) that is often introduced into an aquarium to eat pests such as the glass anemone, *Aiptasia pallida*. The shrimp are rather small and tricky to watch; it takes some degree of stealth and close observation. However, if this is done, you will see that the "beneficial" shrimp are continually picking at the surface of rocks, eating small *Aiptasia*, to be sure, but also nipping the polyps of various prized corals, as well as small snails, small crustaceans, and even the tentacles of larger corals. If aquarists have the patience to sit and watch, and take the time to train themselves to observe, all of this is visible.

REEF PRIMER

Throughout this eye-opening book, Denise and Larry Tackett show the results of this sort of patient observation and inspire all of us to "look closer" at reef life.

This is a volume to whet the appetite of a naturalist—whether a diver, snorkeler, marine aquarist, or armchair adventurer. If I am not mistaken, it will make you want to get to the reef and see if you, too, can witness some of the animals and interactions shown here. It will certainly make you a more informed observer and allow a fuller appreciation of the next reef you visit.

If the authors succeed in simplifying marine biology concepts and basic terminology for more divers and aquarists, they will have done a yeoman's service to all of us who care about the future of these threatened environments. Coral reefs will need all the friends we can muster in the century ahead, and books such as this can help generate them. Full of beautiful and uncommon organisms and important-to-know behaviors, ***Reef Life*** is an obvious labor of love and a highly commendable primer on reef life.

—Ronald L. Shimek
Wilsall, Montana

Ronald L. Shimek is a marine zoologist and former chairman of the Biology Department of the University of Alaska, Anchorage. He is a biology consultant, certified scuba and submersible diver, and author of numerous articles on marine invertebrate life.

Small beauties: a shoal of Blue-green Chromis (*Chromis viridis*) hovers just a dash from the protective branches of a colony of a staghorn (*Acropora* sp.) coral thicket.

Prologue

Reef Exploration

An Introduction to Nature's Most Diverse Ecosystem

"It doesn't matter where on Earth you live, everyone is utterly dependent on the existence of that lovely, living saltwater soup. There's plenty of water in the universe without life, but nowhere is there life without water."

Sylvia Earle,
Change (1995)

OUR LIFE WITH CORAL REEFS BEGAN ALMOST two decades ago when the director of a cancer research group at an American university came through the receiving line at our wedding and whispered, "Think Palau," before disappearing into the crowd.

His group had been working to isolate marine natural products for many years, and all the collecting and research was starting to pay off. New compounds from invertebrates, particularly sponges, were beginning to show potential as anti-cancer agents. The spread of AIDS was intensifying the search for anti-viral compounds. Funding for expeditionary work was increasing.

The offer of a part-time job collecting sponges gave us an opportunity to work with coral reefs that we had never imagined in our wildest dreams. We had accom-

Diversity's vivid realm: myriad feather stars (Crinoidea) and other reef animals festoon a giant sponge (*Agelas* sp.).

1 Feather stars (Crinoidea)
2 Sponge (*Agelas* sp.)
3 Encrusting sponges, algae, sea squirts
4 Sponges (Porifera)
5 Soft coral (cf. *Xenia* or *Heteroxenia*)
6 Torch Coral (*Euphyllia glabrescens*)

panied the university on a collecting expedition to Papua New Guinea the year before, and when we returned from our honeymoon, the director offered us on-going jobs collecting sponges in remote locations. Much to the dismay of our family and friends, we cast our fates to the wind and gave up established careers in banking and chemical engineering. We sold our house, furniture, and car, and put whatever else was left into storage. Then, after packing one suitcase, camping and diving equipment, and cameras, we took off on the adventure of a lifetime.

We had expected the expedition to last a year. In reality, our belongings stayed in storage for more than a dozen years. During that stretch of time, we graduated from one suitcase to over a ton of equipment and logged several thousand dives in dream destinations in the Western Pacific and Indian Oceans. Our parents just about gave up hope that we would one day return permanently to the States and lead "normal" lives. They liked to tell people that we left on our honeymoon and didn't come back for 13 years.

As sponge collectors, we followed a simple routine wherever we went in the Indo-Pacific. With help from the local authorities, we chose an expedition site, usually a remote beach, where we set up a base camp and hired a local fishing boat for diving. Our camp was fully self-contained. It had sleeping and storage tents, a kerosene stove and lamp, small generator, scuba compressor, and all the diving and collecting gear we'd likely need. Being the only members of the expedition team, we learned to live in relative isolation from the modern world for six to nine months of the year. Even when we had the luxury of local housing, we often didn't have modern conveniences like electricity, running water, and telephones, although we managed quite well.

Our system was to dive three times a day searching for specimens. We concentrated on sponges because they have unique defensive chemicals that repel predators, and sometimes kill cancer cells, unlike some of the higher invertebrates that rely on external defenses such as spines and shells.

At last report, the university has several compounds in various stages of development and clinical trials. They have isolated a number of promising compounds from marine animals like sea hares, bryozoans, and sponges. These compounds, known as dolastatins, bryostatins, and spongistatins, respectively, represent bright prospects among new anti-cancer substances. They force certain cancers into remission and do so with far fewer side effects than other currently available chemotherapies. Although much work remains to be done, there is considerable optimism regarding these substances. If all continues to go well, therapies based on these compounds could be available in a few years.

As sponge collectors and field workers for 13 years in 11 countries, we accumulated thousands of dives on remote reefs. This gave us an extraordinary opportunity to observe marine life in a variety of locations. While the work was physically demanding and the living conditions primitive, we had the obvious rewards of free diving, foreign travel, and a sense of service to mankind. Over the years, we had time to enrich our own lives and to hone our photographic skills.

With this book, we would like to share with you some of the wondrous secrets of the underwater world that have taken us years to learn. We hope to enrich your knowledge and inspire you to learn more about the lives of coral reef animals. Like us, you may come to see reef life through new eyes. Only when many more of us understand the priceless intricacies and beauty of this underwater world will the future of our world's reefs be in more secure hands. ■

"Critter eyes": nestled in a bed of soft corals, a giant crocodilefish (*Cymbacephalus* sp.) lies in predatory wait, unnoticed by casual passersby. To appreciate the intricate life of a coral reef, one must slow down and learn to look closer.

PATIENCE REWARDED

Most of us are bedazzled by our first look at coral reefs, but altogether too many newcomers to coral reefs end up missing the real show in favor of watching a few of the bigger or more menacing stars. Divers, snorkelers, and even marine aquarists are often so intent on seeing the large, spectacular fishes that they miss much of the delicate beauty, the hidden animals, the astonishing behaviors and bizarre life forms right in front of them.

How to become a better observer? My advice to all divers with a curiosity about natural history is to work on developing **"critter eyes"**—the ability to spot things that may not want to be seen. The best way to develop critter eyes is simply to slow down and observe what is in front of you.

Some people seem to dive just to see how much territory they can cover. I urge them to stop and concentrate on one small area, perhaps the size of a basketball. It takes three to five minutes for the animals to get used to a new presence, after which they should return to their normal activities, provided one assumes the role of passive observer and do not become an intruder by poking, moving, or manipulating any animals or their surroundings. I like to build up a trust with my subjects—they need to know that I will back off at any time if they seem disturbed.

Sometimes the best way to find a perfectly camouflaged organism is to look for movement. Some animals are so well hidden that you cannot rely on your eyes alone to recognize them. Instead, look for the flutter of a gill cover, the flick of a tail, the movement of an eyeball, the soft rays of a fin. Look for something out of place, something different. If you find a new animal, watch what it does. Does it have legs, fins, tentacles? Does it have eyes, a mouth, none of the above? How do nearby animals react to it?

Marine aquarist friends tell us these rules apply to watching the life in a captive reef: slow down, spend time watching a narrow field of view, and the things you see may amaze you. ■

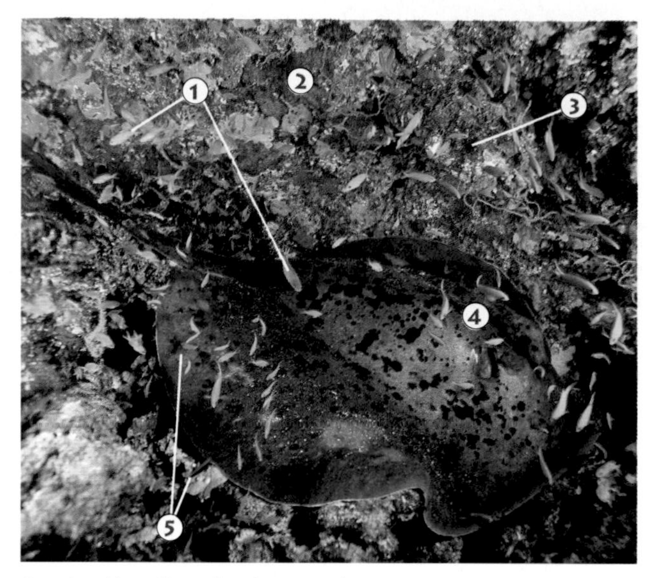

1 Anthias (*Pseudanthias* spp.)
2 Sponges (Porifera)
3 Coralline algae
4 Blackspotted Stingray (*Taeniura meyeni*)
5 Bluestreak Cleaner Wrasses (*Labroides dimidiatus*)

Intertwined lives: colorful anthias and a Blackspotted Stingray (*Taeniura meyeni*) pause at a cleaning station, where various fishes come to be groomed by Bluestreak Cleaner Wrasses (*Labroides dimidiatus*) that remove attached parasites and damaged or diseased tissue.

An eye for the beholder: an extreme closeup catches a reflected world in the eye of a frogfish (*Antennarius* sp.). All too many reef visitors intent on seeing big pelagic animals would never have noticed the fish, let alone its detailed beauty.

■ THE PHOTOGRAPHS

The photographs in this book were all taken *in situ*, using a variety of camera systems.

Wide-angle shots were taken with a Nikonos III camera and 15 mm lens, Nikonos RS with 28 mm lens, or with various Nikkor wide-angle lenses on Nikon F-3 or F-4 camera bodies in housings.

Macro shots were taken with Nikkor 60 mm or 105 mm lenses on Nikon F-3 or F-4 camera bodies in housings. We also used the Nikonos RS with a 50 mm lens for macro work.

For **extreme close-ups**, we used Nikon 3T and 4T diopters and a Kenko 2X teleconverter, alone or in any combination with the 105 mm lens.

Most shots were taken with dual strobes. Our preferred setup, and the one we use exclusively now, is the Nikon F-4 camera in a Nexus F-4 Pro housing with dual Ikelite 200 strobes on Ultralight strobe arms.

We prefer Velvia or Ektachrome VS film for most underwater work. We use Fujichrome 100 and Kodak Ektachrome for wide-angle work.

Patience is the one item the underwater photographer cannot purchase, and it is at least as important as all others. As in simply finding the less obvious animals, it is imperative to slow down and observe. Don't despair if a newfound fish or other animal moves away as soon as you approach or attempt to get closer. Many animals are creatures of habit and will return to the same spot time and time again. This bit of knowledge is particularly useful when photographing marine subjects because it allows you to move into position before the animal returns to its anticipated spot.

Keeping a **reef log** of what you observe, as well as the time and place, can often help in predicting the next appearance of some animals. Certain behaviors are linked to time of day and/or lunar cycle. An awareness of these connections can make underwater photography sessions more productive and rewarding. ■

Life in miniature: a tiny goby (Gobiidae) perches on a tunicate or sea squirt (Pycnoclavellidae) on a reef off Komodo Island, Indonesia. Coral reefs are home to all but a few of the world's approximately 35 known phyla and some of the most densely concentrated and colorful life forms on Earth. The advent of scuba and snorkel gear has made coral reefs readily accessible to naturalists—professionals and amateurs alike—drawn by the greatest diversity of easily observable life on our planet.

1 Orange cup corals (*Tubastraea* sp.)
2 Feather stars (Crinoidea)
3 Encrusting sponges (Porifera)
4 Colonial tunicates (Ascideacea)

Nocturnal reef: the blazing colors and shapes of a coral outcropping on a Komodo Island, Indonesia, dive site known as "Cannibal Rock" suggest an artistically landscaped garden. In truth, it meets Ronald Shimek's description of the coral reef as "a wall of mouths," made up of carnivorous night-feeding cup corals, feather stars, filter-feeding sponges, and tunicates, all waiting to catch passing plankton and other particulate matter.

Armed hunter: a lone yellow crinoid (Crinoidea) spreads its feathery arms to catch plankton carried in the water column over this reef in the Sangihe Islands, Indonesia. Flowerlike, the crinoid is an echinoderm and related to the sea stars.

■ REEF RANKINGS

There is currently no universally accepted inventory of coral reefs in the world, but the relative extent of reef development in various countries is shown in the following charts. These measurements are generally considered low, and are taken from satellite imagery that can measure shallow-water reef structures only. Other estimates place the world total at more than 600,000 square kilometers or about 232,000 square miles—less than one percent of the area of the world's seas.

Coral Reefs by Region

Reef area in square kilometers

Pacific	108,000	
Southeast Asia	68,100	
Indian Ocean	36,100	
Caribbean	20,000	
Middle East	20,000	(including Red Sea)
Atlantic	3,100	(excluding Caribbean)
Global Total	**255,300**	(98,571 square miles)

Selected Coral Reefs by Country & Geographical Grouping (*Reef area in sq. km*)

Australia	48,000
Indonesia	42,000
Philippines	13,000
Papua New Guinea	12,000
Fiji	10,000
Maldives	9,000
Saudi Arabia	7,000
French Polynesia	6,000
India	6,000
Marshall Islands	6,000
New Caledonia	6,000
Solomon Islands	6,000
Lesser Antilles	1,500
U.S.—Hawaii	1,200

Reef estimates based on World Conservation Monitoring Centre dataset "Shallow Coral Reefs of the World," and "New Estimates of Global and Regional Coral Reef Areas" by Mark Spalding and A.M. Grenfell, *Coral Reefs* (1997) 16:225-230. ■

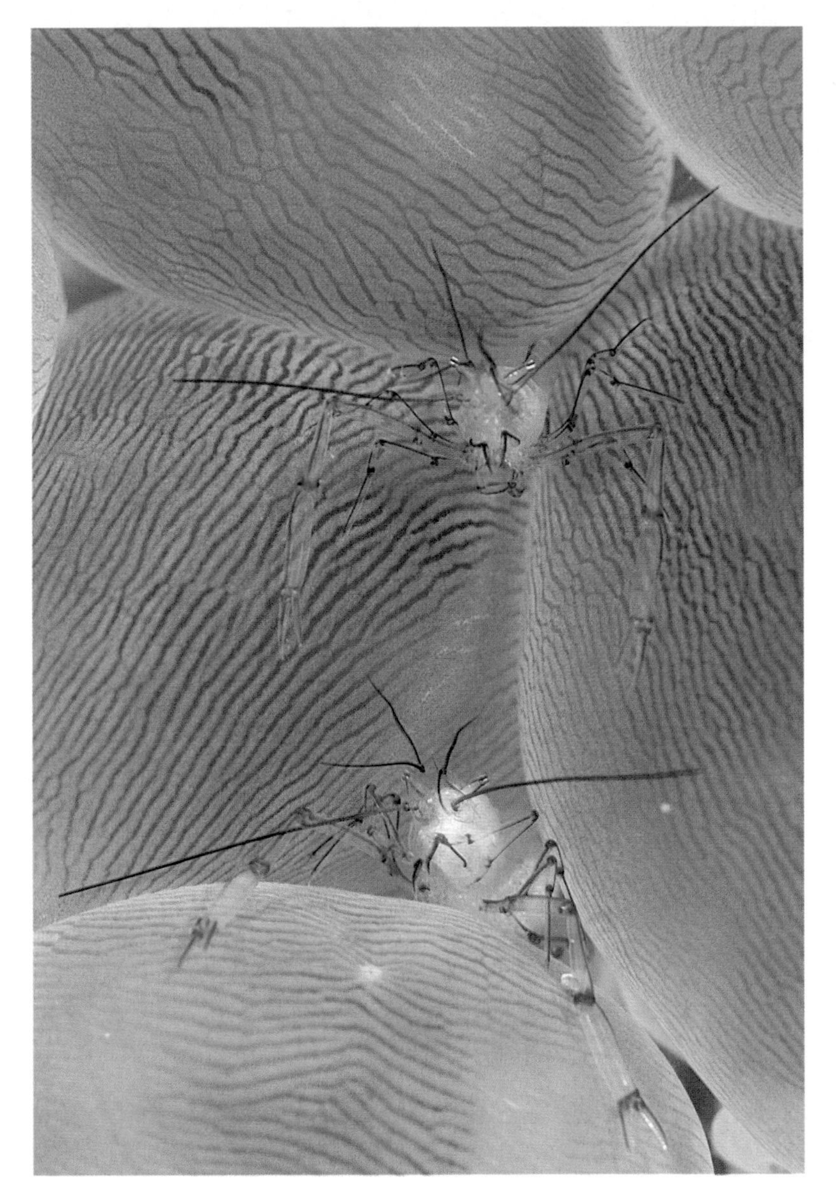

Delicate beauty: a pair of translucent coral shrimp (*Vir philippinensis*) live among the turgid polyp vesicles of a bubble coral (*Plerogyra sinuosa*). The shrimp are considered obligate commensals—found only with this particular type of coral and seemingly unable to survive without its protection. A place of intense competition and fierce predatory pressures, the coral reef remains a largely unexplored wilderness with countless new species and animal relationships waiting to be discovered.

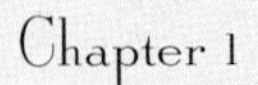

Chapter 1

Reef Origins

Oases in a Watery Desert

"In the biting honesty of salt, the sea makes her secrets known to those who care to listen."

Sandra Benitez,
A Place Where the Sea Remembers (1993)

My first impression was of having landed in a scene of utter destruction or perhaps a place that had never seen life. The underwater site was strewn with cracked boulders that might have been cast down in fury from the heavens. Rusty brown deposits of iron oxide had settled on every exposed surface and filled every crack. Trapped gases escaped fissures in the massive rocks, sending bits of debris swirling, and bubbled slowly toward the surface. No living thing came into view to lend a sense of earthliness to the scene. We could have been on Mars, so otherworldly was the setting. In reality, we were diving on an underwater volcano in a remote area of Indonesia. As I explored further, my eyes gradually became accustomed to the monochromatic scene and began searching the sediment for traces of the reef organisms that usually proliferate in such shallow waters. I didn't have to go far. Amid this desolation, the first signs of stirring life were everywhere—a variety of algae, first links in the ocean's food chains, were thriving under a layer of volcanic detritus. On the algae were tiny sea stars, nudibranchs, worms, and other invertebrates. Random clumps of sponges and corals managed to survive despite the adverse conditions. We

Future reef: barren sulphur deposits rest on bubbling underwater volcanic vents in the Sangihe Islands, Indonesia.

Lava rivers: black, hardened lava flows mark Ruang Island in the remote Sangihe Islands, Indonesia. While initially destructive, old lava flows provide a perfect substrate for coral reef formation in warm tropical waters.

were witnessing a new reef in the making—life forms assembling themselves in ways that have astonished and awed humans throughout history.

TO ANCIENT MARINERS, CORAL REEFS WERE deadly hazards where ships foundered and souls were lost. To tropical islanders, reefs are life itself, giving safe harbor from the violent seas and yielding a ready harvest of fishes, crustaceans, mollusks, and other readily caught and collected foods and materials. To early scientists, corals and the reefs they formed were a profound bafflement.

Thriving only in warm, clear tropical waters, reefs and their proliferation of life forms seem to defy the usual rules of nature. In other realms, plants and animals congregate in areas with obvious sources of fertility, such as good soils, nutrient-rich rivers and streams, and adequate light. In the shallow, pristine waters where coral reefs occur, nutrient levels are so low as to be virtually undetectable. How can this be?

For centuries, humankind grappled with this paradox. Corals were initially classified as minerals, because of their stony skeletons, then plants, and finally as the animals that they are. It was the discovery in 1881 of **zooxanthellae**, symbiotic algae cells found living in the flesh of corals and other reef animals, that led to the ongoing unravelling of the mystery.

Reduced to elemental simplicity, the single-celled algae proliferate in the sun-drenched conditions of the tropical shallows, deriving their primary energy from solar radiation and finding protective shelter in the tissues of various reef animals. Synthesizing an excess of energy-rich products, the algae ooze nutrients into the flesh of their host corals, which use this energy to drive a series of complex reactions in which calcium is extracted from seawater and deposited as hardened coral skeletons.

What exactly is a coral reef? To the geologist, a coral

REEF FORMATION

Setting out to unravel the mysteries of coral reef growth, Charles Darwin set out on the good ship *Beagle*, literally to plumb the depths of the tropical seas. His findings, published in 1842 as *The Structure and Distribution of Coral Reefs,* established for the first time a scientific explanation for how coral reefs came to be.

Darwin split all coral reef systems into three major types: fringing reefs, barrier reefs, and atolls. A **fringing reef** is a mass of coral growing in shallow waters adjacent to the shore of a continental land mass or an island. A **barrier reef** also lies parallel to a shoreline, but is separated from the land mass by a distinct lagoon or channel. An **atoll** is a ring or partial ring of coral topped by a sandy island that surrounds an interior lagoon.

Darwin's theory, radical at the time, stated that the fringing reef is the basis for all types of coral reefs. It forms in the sunny shallows along a shoreline, but in some cases the land mass subsides or sinks deeper into the sea. Darwin believed that, while the waters rose, the living corals were able to grow rapidly enough to survive, living atop a slowly sinking mound of their dead coral ancestors. In time, the coral mass appeared to be an elongate barrier reef, well separated from shore, perhaps surrounding the tip of a volcano that had settled in the depths.

The theory was tested in 1904 by a 330 m (1,082 ft.) core sample drilled on Funafuti Atoll in the South Pacific. It clearly showed layer upon layer of shallow-water corals. Darwin was right. With minor revision, his concepts about reef formation have withstood the tests of time and are considered valid today.

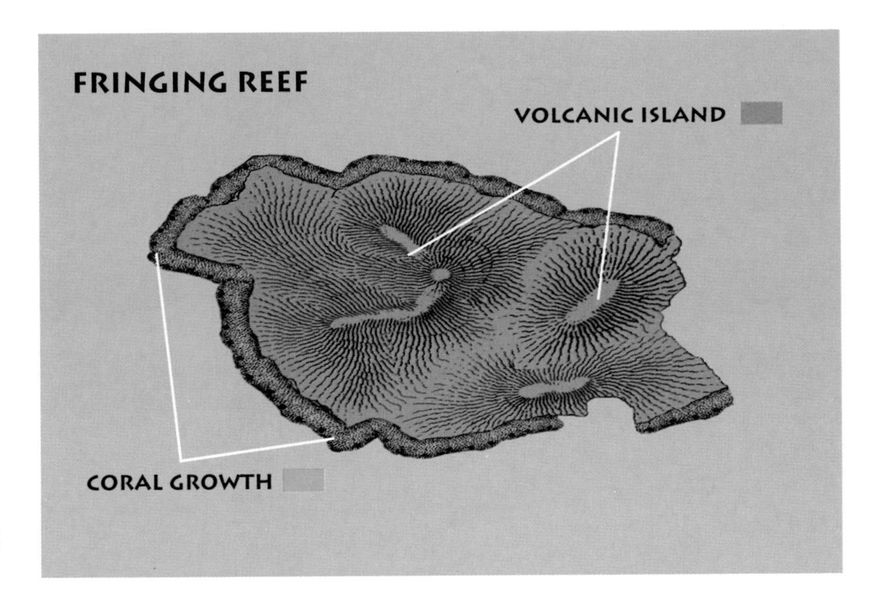

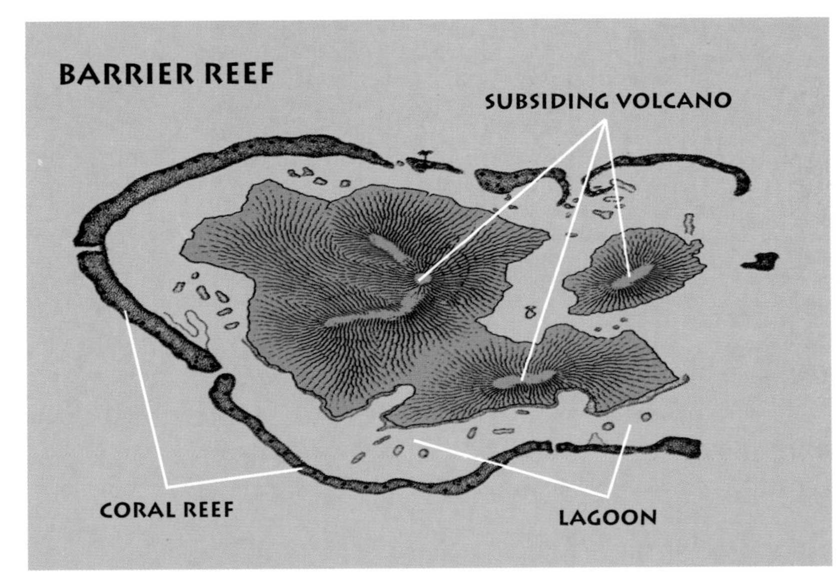

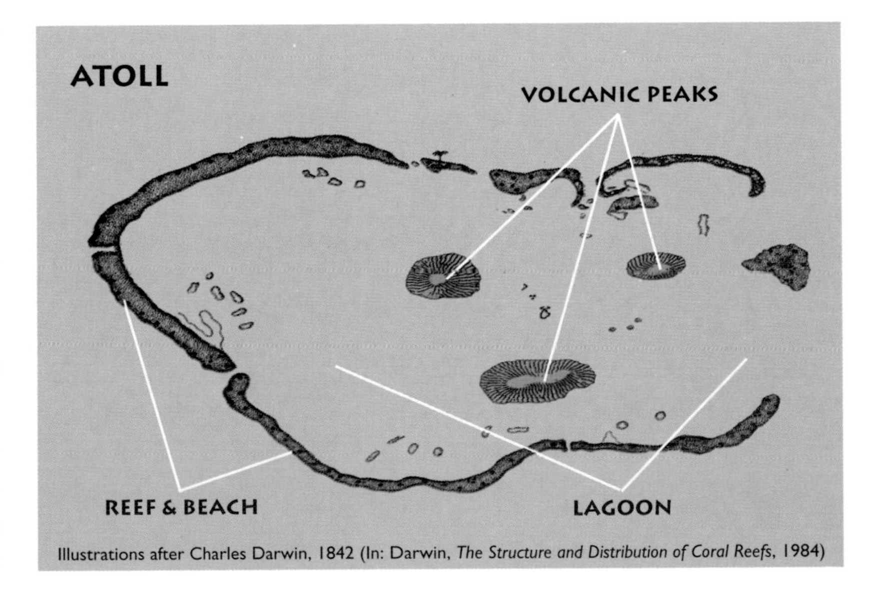

Illustrations after Charles Darwin, 1842 (In: Darwin, *The Structure and Distribution of Coral Reefs*, 1984)

First life: macroalgae begin to cover the barren boulders of an underwater volcano. With the algae in place, other organisms begin to appear.

Sea stars: one of the first colonizers of the new reef is a tiny sea star (Phylum Echinodermata) that settled out of the water column as a microscopic larva.

Bristleworms: a small polychaete worm scavenges through the algae for edible detritus.

Stony corals: a new cup coral (*Tubastraea* sp.) has settled onto the rocky substrate and begins to grow.

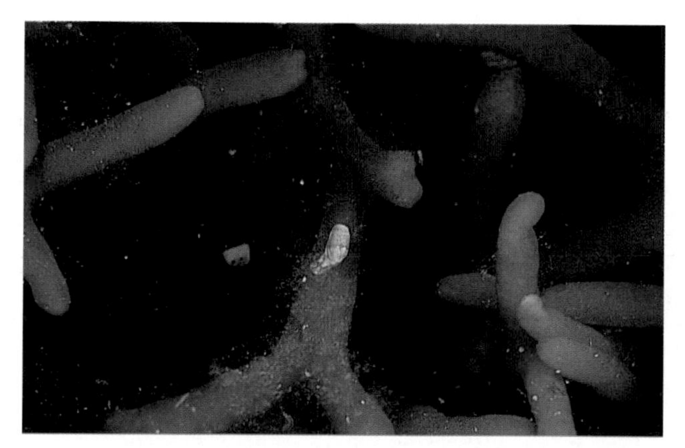

Mollusks: a small bubble shell sea slug (*Bulla* cf. *vernicosa*) grazes on the algae and offers a target for hunting fishes.

Fishes: as a new web of food sources spreads, fishes such as this blenny (*Ecsenius pictus*) soon appear.

reef is a mass or ridge of calcium carbonate that has been built up over time by the secretions of stony corals and other aquatic organisms, including calcareous algae and shelled mollusks. Over time, the rocky foundation of the reef is fused into a stony mass that may be hundreds of feet thick. In the case of the Great Barrier Reef, the biological construction extends some 1,900 km (1,140 mi.) in length with almost 3,000 individual reefs.

A coral reef is not just a heap of inanimate mineral material. It swarms with life and represents one of the most complex ecosystems known, composed of thousands of plants and animals that have evolved intricately intertwined lives in one of the most competitive and species-rich habitats in the world.

As these so-called stony corals grow, many complex processes are put into motion. Habitat is created, attracting all manner of aquatic life with essential shelter from open-water predators. Where these plants and animals gather, there are wastes, which serve to feed other populations of plants and animals. In time, a reef community evolves: a collection of corals, sponges, algae, fishes, and a proliferation of other life forms that exploit every nook and cranny in the reef labyrinth.

Like an oasis in a desert, a coral reef is a fertile hub in the ocean. It is a spawning ground for many fishes and invertebrates and therefore generates tremendous quantities of **plankton**, drifting organisms of various sizes from microscopic algal cells to larval fishes to small invertebrates that are all important food sources for commercially valuable oceanic species.

REEF PARAMETERS

Coral reefs do not pop up just anywhere. As anyone who has traversed sections of the Caribbean or the tropical Indo-Pacific by plane or boat will know, there are vast expanses of warm seawater absolutely devoid of reefs. Some tropical islands are ringed with coral, others are surrounded by deep blue sea. A single island may boast luxurious coral growth along one coast, and nothing but open water off its opposite shore. This is not simply nature acting in a random manner.

Several conditions or environmental parameters are necessary for healthy reef development. Certainly one of the most important requirements for coral growth is **warm water**. Coral reefs are found in areas with average temperatures ranging from 21 to 29 degrees C (69 to 85 degrees F), although they can live at temperatures outside this range. The worldwide average temperature on a coral reef is 27.1 degrees C (81 .7 degrees F). Corals can become stressed and die if temperatures vary much above or below the normal levels for their location. Ideal temperatures are found between 30-degrees North latitude and 30-degrees South, a band roughly 7,040 km (4,400 mi.) wide encircling the globe.

Excellent water conditions are also a must for coral growth. Constant surges from wave action provide vital **oxygenation** of the water, prevent overheating and stagnation, and flush wastes from the surface of sessile life. **Water clarity** is essential, because solar rays must be able to penetrate the water and sustain the vitality of zooxanthellae. Turbid conditions block sunlight and are often coupled with the presence of runoff from terrestrial sources or pollution. Sediment flowing into the sea from rivers, streams, or sites of erosion all work against coral health and the existence of reefs. Nutrients, such as wastes from development or agriculture, can spur the proliferation of algae that outcompetes and overtakes many reef-building corals. Most corals are sensitive to chemical toxins, and few reefs are able to thrive near areas of industrial development or commercial shipping. Reef waters are exceptionally clean, with a pH of 8.2 and virtually undetectable levels of such nutrients as nitrate and phosphate that are common in areas rich with organic wastes.

Constant levels of **salinity** or salt content, typically 35 to 37 parts per thousand (specific gravity of 1.024 to 1.027), depending on the reef and the water temperature) are vitally important, and areas subjected to wide seasonal variations because of currents or runoff from floods are poorly suited to reef organisms.

Finally, reef-building corals need a firm footing in relatively **shallow water**—generally no deeper than 30 m (100 ft.)—to establish themselves. The majority of reef systems are found anchored on continental shelves (shelf reefs) or out in deeper water where they grow in association with oceanic islands.

Perfect conditions for corals are far from uniform in the world's tropical seas. Coral reefs are estimated to cover less than one quarter of one percent of the total marine environment on earth. Shallow reefs, as mapped by satellite, total some 255,000 km^2 (98,450 mi.^{2})—an area not much larger than Great Britain. ■

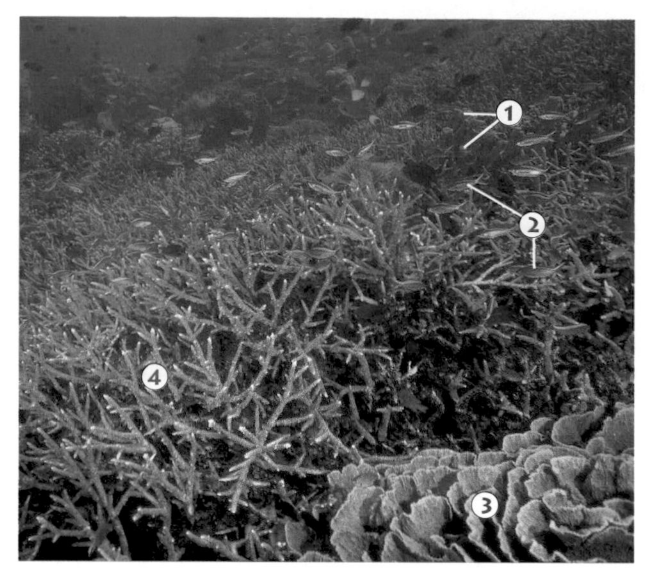

1 Tube sponges
2 Yellowstripe or Purple Anthias (*Pseudanthias tuka*)
3 Plate coral (*Montipora* sp.)
4 Staghorn coral (*Acropora* sp.)

THREE REALMS

In order to understand a coral reef, it is helpful to think of it as three groups, or realms, of organisms.

Reef-builders are the plants and animals that build and cement the structure of the reef. Stony corals and calcareous algae are the key to building a hard, intricate reef structure that shelters the entire community. Sponges, soft corals, and other encrusting organisms stabilize loose substrate by binding stony fragments together, thus providing habitats for fishes and invertebrates.

Cryptofauna often go unnoticed because they are small or hidden, but they riddle the reef and permeate the areas of sand and soft substrate. These crustaceans, mollusks, worms, and others play vital roles in recycling wastes and serving as food for the larger animals.

Free-swimming organisms are the most obvious and include the many fishes and other animals that swim over and around the reef.

The lives and success of each group is dependent on the others, and a deficiency in any one area can seriously disrupt the balance for the other two.

Reef on lava: thriving coral reef, dominated by fast-growing stony corals, covers a bank of lava just a matter of 50 years after a devastating volcanic eruption in the same Indonesian region shown on pages 28-29.

Algal ridge: pounded by the incessant forces of incoming waves and subjected to the furious energy of tropical storms, this breaker zone in the Maldives is dominated by tough calcareous algae with strong-swimming fishes and invertebrate life that can cling to the substrate and maintain itself in protected niches.

Reef crest: forming along the upper edge of the reef, where the reef flat and seaward slope meet, this turbulent zone in Sipidan is populated by fishes that thrive in an oxygen-rich environment and stony, reef-building corals that can endure strong wave action. An algal ridge may form here in especially heavy surf conditions.

Upper fore reef: typically a long slope from the reef crest to the seafloor, the fore reef is a series of zones whose characteristics are shaped by wave action and the amount of sunlight they receive. Massive reef-building corals, such as this colony of mound-forming *Porites* sp. (**1**), dominate portions of this upper slope in the Maldives.

Mid fore-reef slope: below the most turbulent conditions, colorful soft corals sway back and forth in the currents, trapping phytoplankton—microscopic bodies of single-celled algae drifting in the water column. Also common in the less energetic zones are calcareous macroalgae, sponges, and a plethora of smaller fishes.

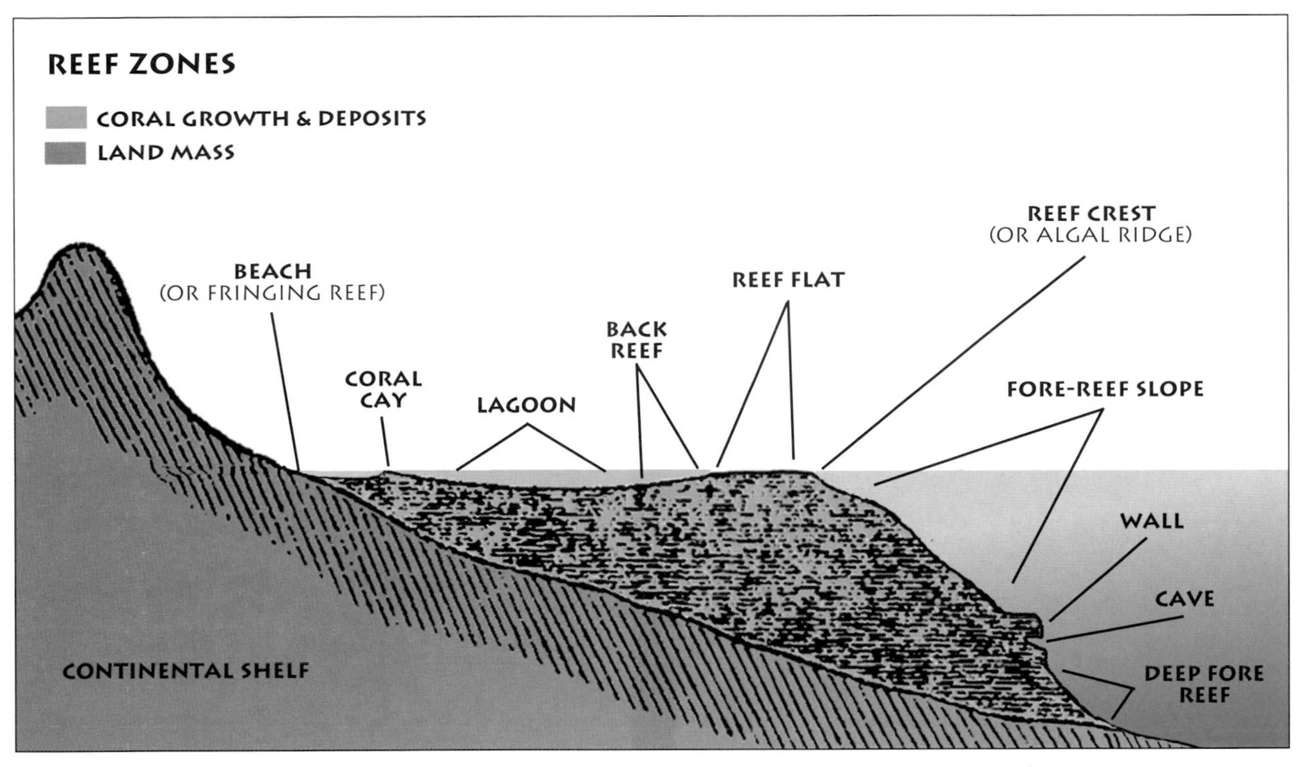

Cross section: with coral deposits shown in pink, this schematic of a typical continental shelf reef displays the major zones, each with its own distinctive assemblages of life forms. Most living organisms are concentrated in the reef crust.

Deep reef: able to thrive in the absence of strong sunlight, sponges and gorgonians (Melithaeidae) dominate a patch of deep fore reef at a depth of 30 m (100 ft.).

Deep lagoon: these giant sponges and colorful crinoids, or feather stars, are filter feeders that often inhabit deeper quiet zones with weak, filtered sunlight.

Atoll: a ring of coral reef and sand marks the spot where a volcanic peak has subsided into the Indian Ocean. This is one of the many faros (ringlike reefs inside atolls) found among the 26 atolls and 1,190 small islands of the Maldives.

Patch reef: submerged islands of coral often dot lagoons, and offer important habitat for many fishes and invertebrates. This patch includes staghorn (**1**) and table (**2**) *Acropora* colonies, among other stony corals.

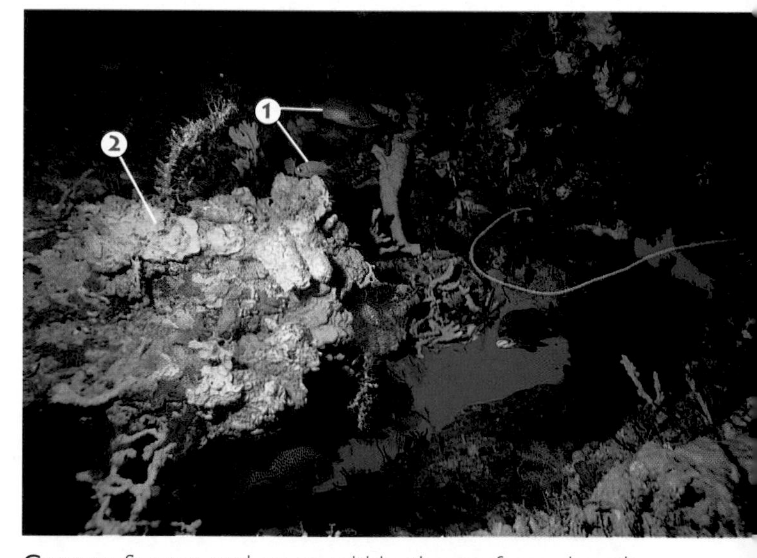

Cavern: fissures and caves within the reef are densely populated with secretive fishes (**1**), sponges (**2**), and other organisms that have adapted to low levels of light and relatively still conditions.

Wall: Scarletfin Squirrelfish (*Sargocentron spiniferum*) shelter among various gorgonians and stony corals clinging to a wall or steep section of fore-reef slope in the Maldives.

Wall hanger: a brilliantly hued tree coral (Nephtheidae) lives attached to the fore-reef, where it is washed with currents bearing the small plankton that make up its diet.

Rocky crevice: Slender Sweepers and Silversides pack a narrow canyon in an Indonesian reef. These niches and grooves in the reef teem with life that finds protection and a constant exchange of water and nutrients.

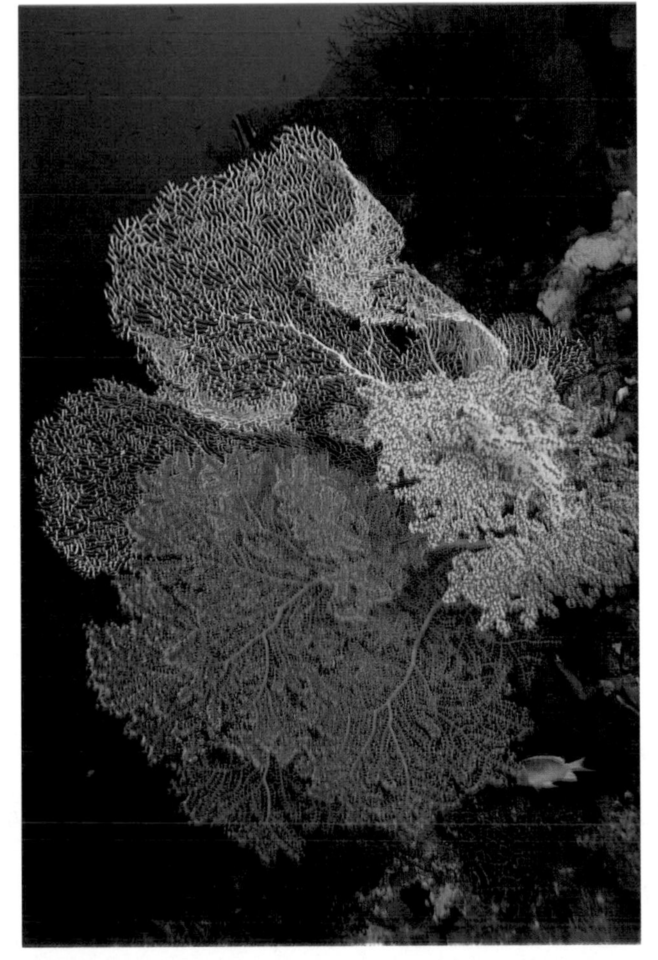

Wall fans: corals with a flexible horny skeleton, these large gorgonians hang from a vertical face on a deep reef in Sulawesi and are perfectly equipped to sieve particulate matter and plankton from the water.

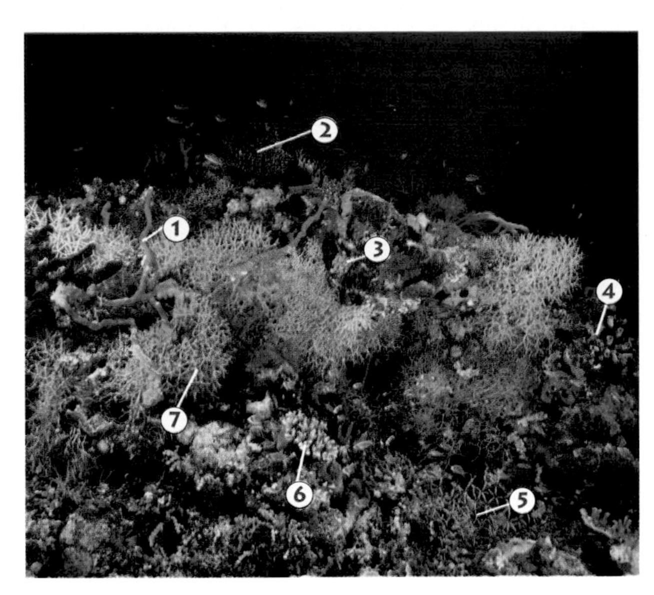

1 Rope sponge (*Amphimedon* sp.)
2 Gorgonian (*Clathraria* sp.)
3 Cup coral (*Tubastraea* sp.)
4 Cup corals (Family Dendrophylliidae)
5 Gorgonian (*Clathraria* sp.)
6 Stony coral (*Acropora* sp.)
7 Gorgonian (*Acabaria* sp.)

COUNTING SPECIES

No one knows for sure how many distinct species exist on the world's coral reefs, and the numbers are increasing as reef exploration continues. Dr. Marjorie Reaka-Kudla has advanced a calculated estimate of between 1 million and 9 million species.

Known species of cnidarians—corals, anemones, jellyfishes, hydroids, and sea pens—number about 9,000, with some 800 species classified as reef-builders. Described and named fish species that commonly associate with coral reefs currently total about 7,000.

Vast reef areas of the world—up to 90 percent of the Indo-Pacific—are still unstudied by biologists. ■

Observable diversity: coral reefs—lacking insects—do not quite match the rainforests for species richness, but far exceed them in accessibility to human observers, who can approach them with ease using scuba or snorkel gear. Landlocked naturalists are now keeping miniature-scale reefs, replete with reef-building corals, in marine aquariums that allow year-round viewing. One reef aquarium system at the Smithsonian Institution, stocked originally with coral rock and sand from the Caribbean Sea, was found to have more than 6,000 different species living in residence.

Ascidian garden: in a miniature bed of glorious color, several species of tunicates or sea squirts (Ascidiacea) vie for growing room on coral rubble. Some sea squirts form small colonies (**1**), while others may form mats of thousands of bodies called zooids (**2**). These water-filled animals tend to collapse temporarily if touched.

Live coral caves: coral hermit crabs (*Paguritta* sp.)(**1**) face out of the burrows vacated by tubeworms in a live colony of stony coral (*Montipora* sp.). Numerous invertebrates find long-term shelter within colonies of coral. A single large head of stony coral may contain thousands of worms, crabs, shrimps, and small fishes.

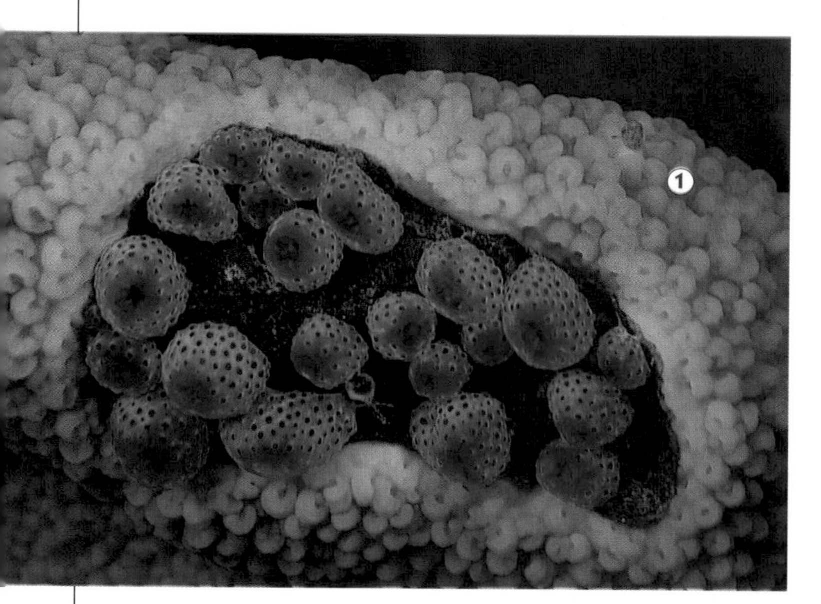

Island of opportunity: sea squirts colonize a patch of bare coral skeleton, surrounded by the live corallites of stony coral (*Acropora* sp.) (**1**). Every possible piece of exposed rock or bare coral skeleton offers a settling area for larval corals, sponges, tunicates, and myriad other organisms that must have a foothold to establish themselves and survive. The tunicates may exude toxins to retard coral growth.

Face off: two species of coral—moon coral (*Diploastrea* sp.), (**1**) and fire coral (*Millepora* sp.) (**2**)—engage in a turf war that will likely end in death for one of the colonies. Corals employ a variety of tactics to defeat competitors, including chemical warfare, the use of powerful stinging tentacles, and simply overshadowing or aggressively overgrowing weaker neighbors.

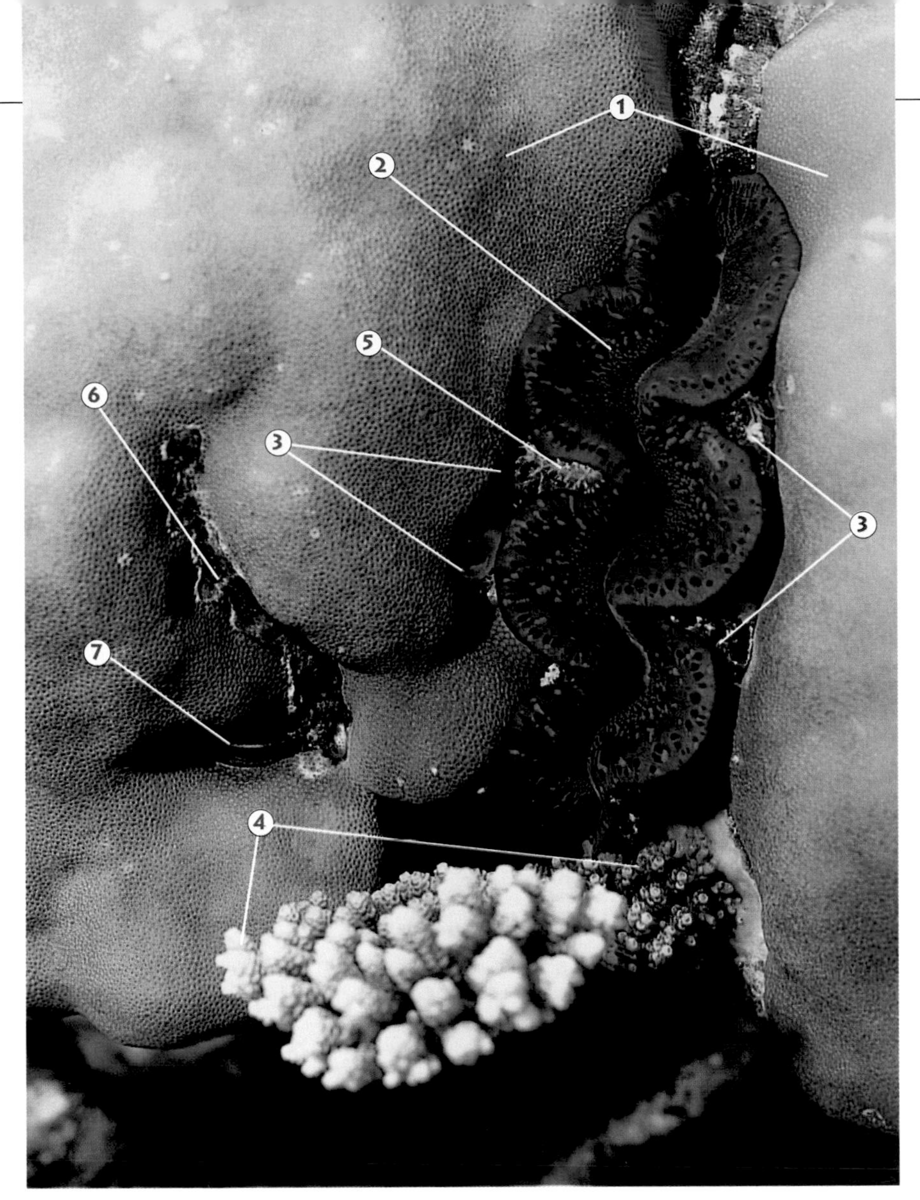

Filling the cracks: various animals crowd into a crevice within a colony of mounding *Porites* coral(**1**), including a burrowing clam (*Tridacna crocea*) (**2**), various solitary and colonial tunicates (**3**), and a small colony of *Acropora* coral (**4**), a small purple sponge (**5**), a patch of boring sponge (**6**), and a juvenile Coral Clam (*Pedum spondyloideum*) (**7**) with a blue mantle and tiny red eyes. Careful examination would reveal more organisms within this microhabitat.

■ COMPETITION FOR SPACE

First-time visitors to coral reefs are often bewildered by the sight of so many different living things crowded into one area and tend to see one huge living mass rather than the individual organisms.

For plants and animals, competition for space is a prerequisite of survival. To ensure their survival, tiny larval fishes and invertebrates must secure a safe spot on the reef when they drop out of the drifting plankton to take up permanent residence.

Prime locations are in high demand, and the great majority of larval animals never succeed in finding a toehold on a reef. The best substrates are usually heavily covered with a variety of life, all of it working daily to maintain its spot in the sun—or in a cryptic, shaded recess of the reef, as the case may be.

Curiously, huge **monotypic** (one species) stands of growth are relatively rare in reef environments. More typical are complex sharing arrangements, in which numerous life forms crowd together and manage to coexist. Sometimes their relationships are harmonious—or at least mutually tolerant. In other cases, a constant state of war, with chemical and physical weaponry, is waged until one animal retreats or perishes. ■

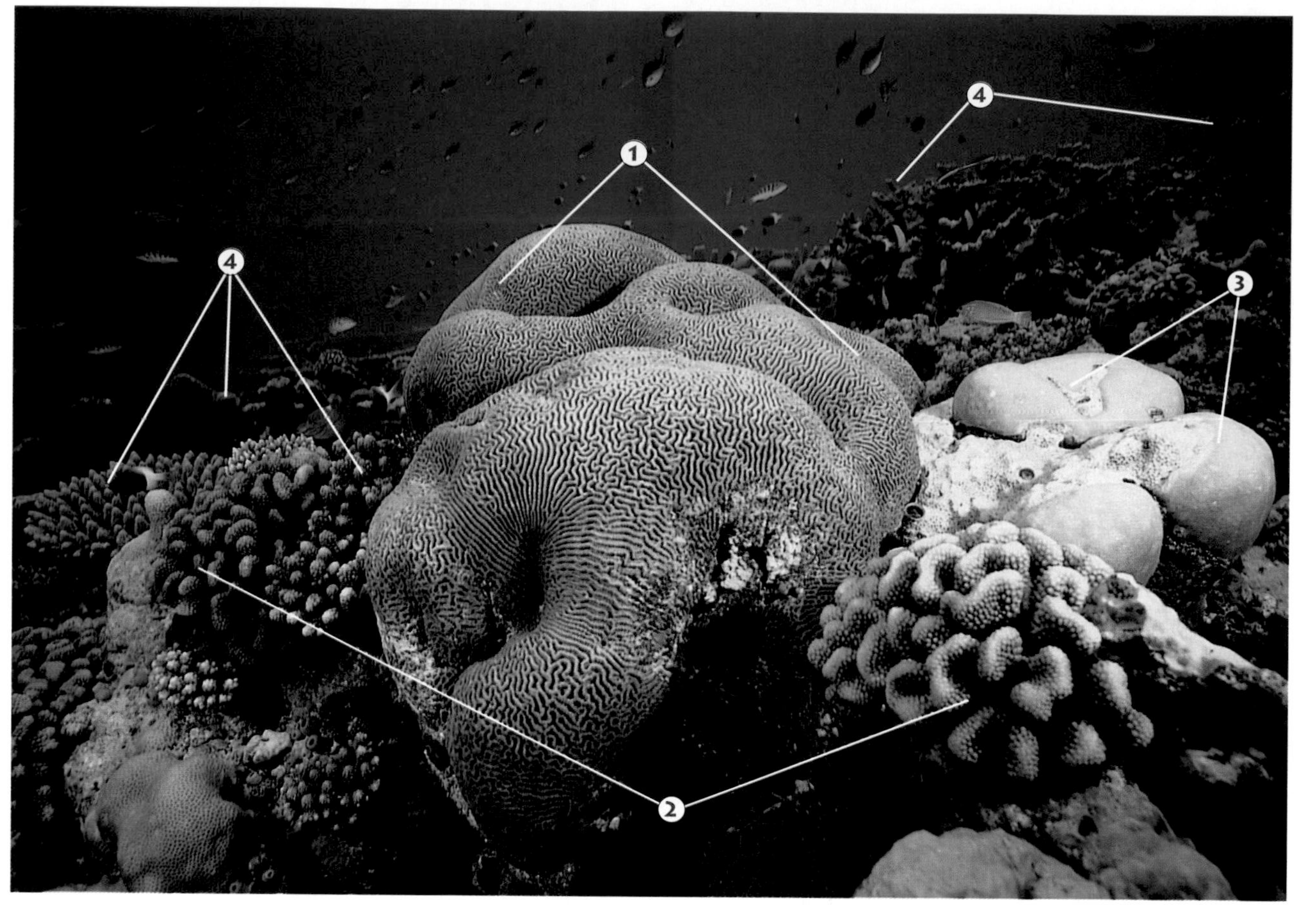

Stony corals: members of a diverse order of corals, the scleractinians are prodigious reef builders. Among the stony corals in this aquascape are: a massive *Platygyra* sp. (**1**), *Pocillopora* sp. (**2**), *Porites* sp. (**3**) and many *Acropora* spp. (**4**).

REEF BUILDERS

Stony corals, known as **scleractinians**, are arguably the world's greatest bio-architects, having assembled the largest structures built by animals on Earth. A reef can be viewed as massive geological structures—underwater rocky ridges of calcium carbonate—or as fragile living ecosystems forming a thin veneer of life over a stony foundation of their own making.

Coral reefs are among the most productive ecosystems in the world, generating tremendous quantities of biomass. A healthy reef can add an estimated 12 to 24 tons of limestone per acre per year to its mass.

Reefs, however, could not exist without the presence of tiny, single-celled organisms known as zooxanthellae. Traditionally considered part of the algae because they contain chlorophyll and conduct photosynthesis, zooxanthellae cells are actually dinoflagellates (having taillike structures that allow them to move) and are now classified as Protists, single-celled organisms that belong to neither the Plant nor Animal Kingdoms. However they are classified, zooxanthellae are the most important primary producers on the reef. Like true plants, they use sunlight and carbon dioxide during photosynthesis to produce oxygen, sugars, and other organic matter. Because the zooxanthellae produce more energy than they need, the coral polyp uses the surplus for its own nourishment and skeletal growth.

The most important of the reef-builders are members of the genus *Acropora*, which include the staghorn, elkhorn, and table corals known to all reef naturalists. The acroporids are fast-growing and opportunistic, spreading and quickly colonizing new areas with relative rapidity.

Other significant reef-building corals include: *Montipora* (velvet corals), *Porites* (jewel corals), *Turbinaria* (cup or scroll corals), *Pocillopora* (bird's nest or cauliflower corals), and *Fungia* (plate or mushroom corals).

Growth forms: stony corals are often identified by their skeletal shapes, including massive, "brain," or boulder (**1**), plating, or foliaceous (**2**), branching (**3**), and encrusting (**4**).

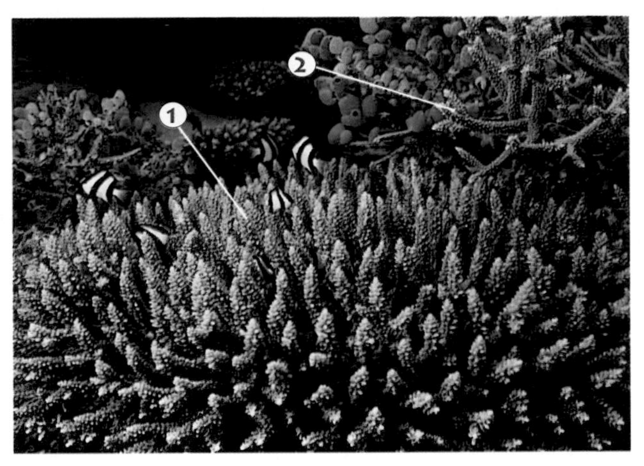

Rampant colonizers: members of the *Acropora* genus are fastest-growing, adaptable, and highly variable in form, from bushy, or caespitose (**1**) to the classic staghorns (**2**)

Table corals: among the most impressive of the corals are the tabular or tabletop species (*Acropora* sp.) that grow so rapidly and large—up to 3 m (9.8 ft.) or more—that they can easily shade and kill other, less dominant coral species.

Columnar corals: in reef zones pounded by waves, stony corals often form very durable knobs or small columns, such as this stand of Star Column Coral (*Pavona clavus*) that demonstrates considerable variability of growth form.

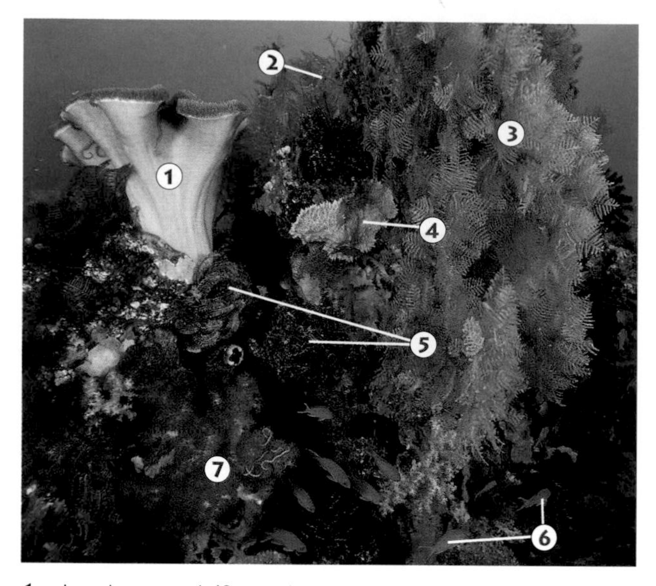

1 Leather coral (*Sarcophyton* sp.)
2 Hydroids (*Lytocarpus* sp.)
3 Hydroids (*Plumularia* sp.)
4 Sponge (cf. *Callyspongia* sp.)
5 Feather stars (Crinoidea)
6 Lyretail Anthias (*Pseudanthias squamipinnis*)
7 Tree coral (*Dendronephthya* sp.)

SPECIES HOTSPOT

Two hundred million years ago, when the Earth's continents were joined together in one large landmass or supercontinent, known as **Pangaea**, there was only one ocean, **Panthalassa**. It was relatively warm, shallow, and supported a wide variety of fishes, soft-bodied invertebrates and large marine reptiles. Pangaea gradually split into seven smaller continents that slowly drifted to their present-day positions.

The landmasses of Asia and Australia and the tens of thousands of islands in between provided the foundation for coral reefs by supplying the essential substrates on which drifting organisms could settle. This region bridged the gap between the Indian and Pacific Oceans, thus allowing organisms from both seas to overlap territories and intermingle, further increasing the species diversity of the area. Today, the region from Southeast Asia to the Great Barrier Reef is the global center of coral reef biological diversity, with the Earth's largest concentration of reef fishes and corals. ■

Diversity galore: Soft corals, sponges, crinoids, hydroids, and a shoal of Lyretail Anthias make up just a small part of this Indo-Pacific reef community.

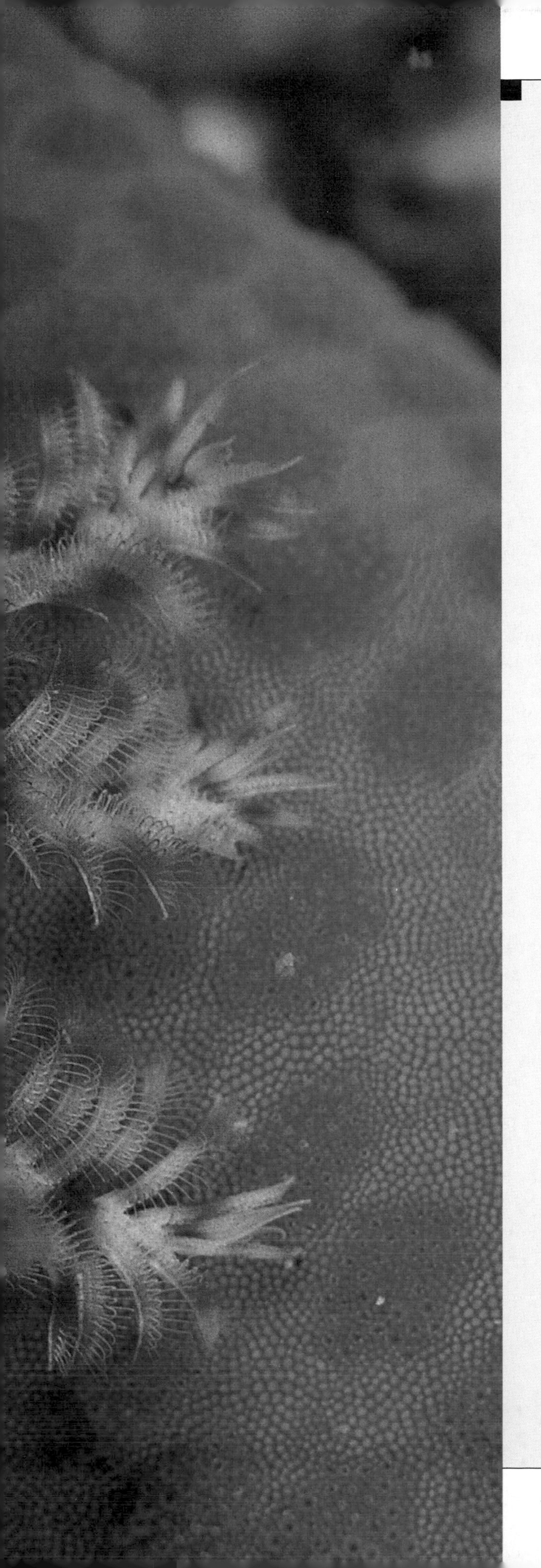

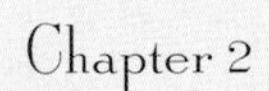

Chapter 2

Reef Dwellers

Diversity's Epicenter: The Plants and Animals of the Coral Reef

"All know that the drop merges into the ocean, but few know that the ocean merges into the drop."
Kabir

Tales of tropical sea snakes and blue-ringed octopuses with toxins powerful enough to kill a man in minutes had filled my head and kept me on my guard since we'd left home to go diving in the Coral Sea in southeast Papua New Guinea. This was our first underwater experience in the Indo-Pacific region, and our first month flew by, offering encounters with countless new and wonderful creatures. Right from the start, we'd been overwhelmed by the tremendous variety of life and color on the reefs. Although relieved not to have had any encounters with the deadly animals we had studied, I was a bit disheartened by the lack of opportunity to photograph such celebrated creatures. At the end of the last dive of the last day, I reluctantly began my ascent. I had reached the surface and already signaled for the boat before I noticed something was wrapped around my arm. For an

Species upon species: biodiversity is evident everywhere on coral reefs, which rival the tropical rainforests in species density. This tiny pink feather star, or crinoid (Crinoidea) with its feathery filter-feeding appendages rests on the arm of a Blue Sea Star (*Linckia laevigata*).

*instant that seemed like an eternity, I found myself eye to eye with a 2 m (6.6 ft.) Olive Sea Snake (*Aipysurus laevis*). That moment of mutual awareness sent both me and the snake into action. We each immediately took off—the greenish snake making a hasty retreat to the depths and I in hot pursuit. The threat of danger never occurred to me. I wanted that photograph. The swimming prowess of the sea snake far exceeded my own aquatic abilities, and I had to settle for a grainy photograph of its tail. That was many years ago, but I still have that shot, and whenever I see it, I smile and recall the time I shared a surprise in the blink of an eye with a sea snake.*

COMPARED TO SO MANY TERRESTRIAL AND marine environments, a healthy coral reef is a hotbed of activity. Everywhere you look, something is happening: animals hunting, burrowing, defending territories, stalking prey, nest building, searching for the perfect mate.

While most of us take this in and come away awestruck, others attempt to count, sort, and classify the life forms found on a reef. A current preoccupation of many ecologists is the study of **biogeography**—simply establishing what species appear where. Measures of **diversity** count the number of species in a set area.

A few elementary rules help explain why coral reefs are so **speciose**—dense with species—compared with rocky shallow areas in more northerly climates. In general, the higher, constant energy input from the sun in tropical areas tends to allow animals to proliferate in both numbers and in different species. Areas with great **spatial heterogeneity**—different types of sheltering cover, habitat, and ecological niches—also favor species diversity. (Both rainforests and reefs meet these criteria, and they are the most diverse environments on Earth.)

Finally, it helps to have periodic—but not frequent—physical disturbances, such as forest fires or violent storms at sea, to upset the balance of power. Areas that constantly enjoy perfect conditions tend to have certain species become dominant, to the increasing exclusion of less competitive species. On the other hand, areas subject to constant or regular physical upset tend to have lower diversity, because only the most hardy species can survive the regular upheavals. For a reef, the occasional damaging hurricane or monsoon may be a very healthy event, breaking the stranglehold of the highly dominant corals and other species. Big storms can also lead to the dispersal of species carried great distances by unusual currents.

Curiously, while the fauna of Caribbean reefs is actually older than that of the Indo-Pacific, it is sometimes described as relatively **depauperate**—reduced or lacking in species compared to other parts of the tropical world. Compared to the Indo-Pacific, for example, the Caribbean has about one-tenth the number of species of corals. This is attributed to the effects of the last Ice Age and to the fact that hurricanes scour the much smaller Caribbean area year in and year out.

Ecologists separate the major environmentally distinct areas into **biomes,** such as rocky shores, sandy beaches, and mangrove-lined shores. Coral reefs are distinguished as being **photic** (having light, versus the dark deep sea, which is **aphotic**). They are **benthic** or associated with the bottom, rather than **pelagic**, or existing in the water column. Finally, they are categorized with other biomes that are characterized as **high energy**, with tremendous influxes of water in motion—unlike the low-energy salt marshes or mangrove swamps.

Similarly, the live organisms found on and around the coral reef are lumped into groups to make them easier to describe and study.

Benthic organisms are bottom-dwellers, such as sponges, mollusks, crustaceans, and fishes that rest on the substrate, such as the gobies, flounders, and many others.

Pelagic species live in the water column, and are typified by many types of free-roving fishes, such as the barracudas, angelfishes, and many sharks.

Epibenthic animals live in the water column, but typically stay very close to the bottom—generally within 3 m (9.8 ft.). Examples of these substrate-huggers include the moray eels, the damselfishes, and the anthias.

Further divisions of reef life include the organisms that are **sessile**—generally fixed to a substrate or other body and not in the habit of moving from place to place. Marine macroalgae, seagrasses, sponges, and corals, as adults, are all sessile.

Freckled Hawkfish (*Paracirrhites forsteri*), an ambush predator, perches motionless and partially hidden on a large, bright orange sponge, ready to launch its next attack. Behaviors and life histories of many reef animals are intertwined.

Motile organisms are capable of movement from place to place, and include the fishes, echinoderms, most crustaceans, and many others.

Curiously, certain animals thought of as sessile can actually muster shifts of position on the reef, if circumstances demand it. Sea anemones, for example, typically stay fixed in one position for years—perhaps even centuries if certain theories about their longevity are correct. But if local conditions become inhospitable—perhaps a nearby coral is nettling the anemone with its sweeper tentacles—the anemone can haul its muscular column up from the substrate and creep or float to another location.

Likewise, many organisms thought of as sessile adults have gone through a life stage in which their larvae were both pelagic and motile, able to zip or undulate through the plankton before settling on the reef.

Plankton itself—small drifting plants and animals—both fuels the reef with crucial sources of food and ensures its future by spreading gametes, larvae, and juveniles from one reef area to others. Much—but not all—reef-associated plankton is not apparent to the naked eye. The size categories of plankton are:

Ultraplankton (less than 5 microns)
Nanoplankton (5 to 50 microns)
Microplankton (50 to 500 microns)
Mesoplankton (500 microns to 0.5 cm [0.2 in.])
Macroplankton (5,000 microns to 5 cm [2 in.])
Megaloplankton (larger than 5 cm [2 in.])

This is more than academic, as reef biologists are finding that many organisms—from sponges and soft corals to certain fishes—are highly specialized in the size and type of plankton they need to feed and thrive.

As soon becomes obvious to even the casual observer, coral reefs are not just pretty panoramas, and the myriad organisms that populate them are unbelievably diverse. Sometimes referred to as "simple animals," many of them are anything but. The following pages provide an introduction to and overview of the profound diversity of reef life. ■

SESSILE ANIMALS

Although not as immediately eye-catching to new reef visitors, the sessile invertebrates that live attached to the hard substrate are some of the most important members of the reef community.

Sea squirts, barnacles, hydroids, and oysters are examples of sessile filter-feeders that need a place of permanent attachment in an area with good water movement.

In a process called **biological erosion**, some reef dwellers—both sessile and motile animals—are able to bore into the limestone skeleton of live corals as well as fused limestone rock in order to create a place for themselves. The borers use chemical and/or mechanical processes to work their way into corals, rocks, and shells. Some secrete chemicals that dissolve the calcium carbonate layers of corals and mollusk shells. Most boring activities take place in dead coral material, often from the underside of a living coral. Some borers are coral-specific, meaning they only bore into the skeletons of certain species even though there is no longer any living tissue. Boring activities ultimately weaken the structure being bored.

Filamentous algae and microscopic bacteria and fungi can bore their way into corals. Sponges are notorious for their boring activities, which can completely destroy a coral head from the inside out, much the way termites destroy the structure of a house by leaving only a thin coating of paint, while giving the illusion of a complete structure. *Cliona* sp., a common boring sponge, is able to perforate a coral head thoroughly and yet leave the living polyps on the surface intact.

Despite the destructive activities of the bio-eroders, their work is of vital importance to other reef organisms that rapidly colonize any new burrow or patch of dead coral skeleton and fill it with new life. ■

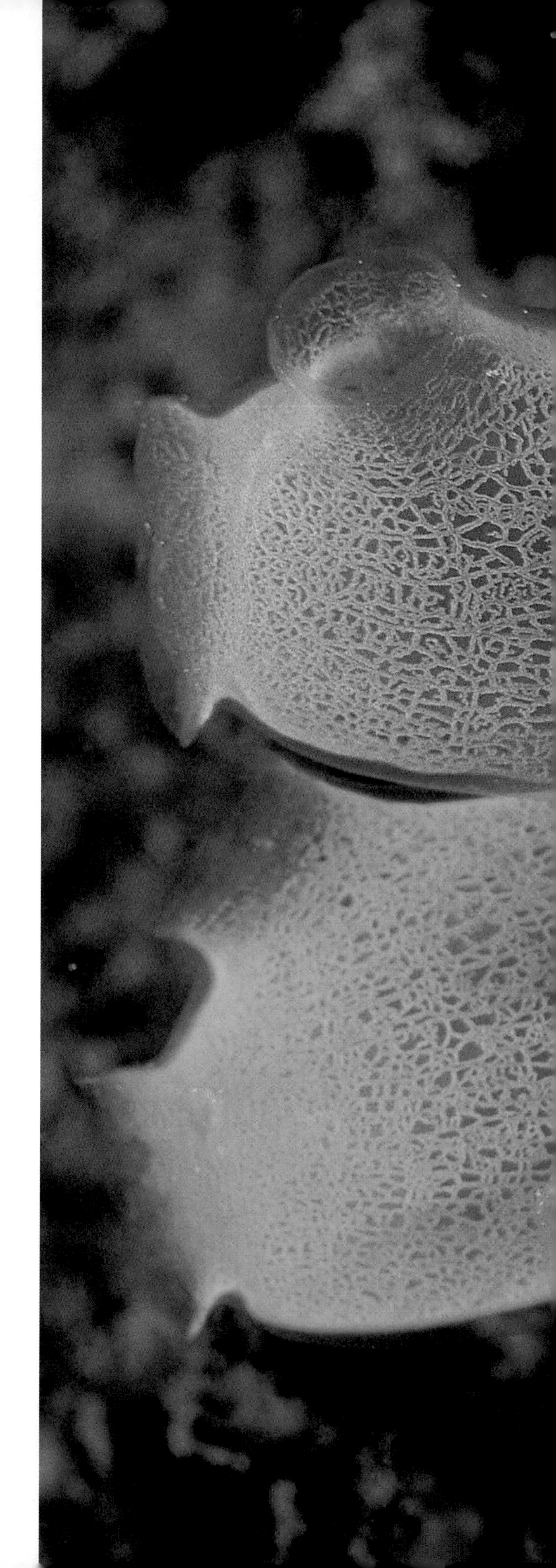

Delicate beauty: colorful, bizarre, and fascinating life forms abound throughout the reef. Here, a cluster of transparent sea squirts, or tunicates (*Rhopalaea* sp.) appear delicate but are successful colonizers of many marine substrates. Often confused with sponges, the sea squirts are actually much more complex animals that have a rapidly swimming larva that looks like a small tadpole. As adults, their only motion is to contract if threatened.

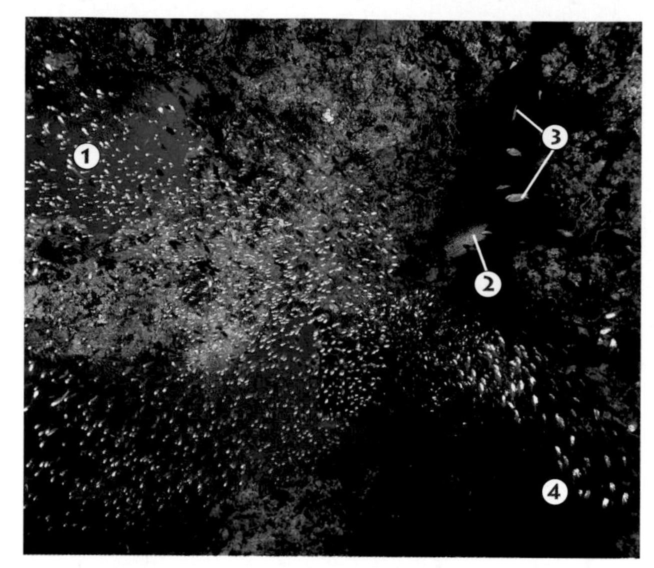

1 Hardyhead Silversides (*Atherinomorus lacunosus*) and Slender Sweepers (*Parapriacanthus ransonneti*)
2 Giant Squirrelfish (*Sargocentron spiniferum*)
3 Lyretail Anthias (*Pseudanthias squamipinnis*)

REEF DIVERSITY

Estimates of the number of living species in each of the major taxonomic groups of marine life vary greatly, and obviously differ significantly from region to region, and certainly from reef to reef. The following general breakdowns give an approximation of the diversity of different phyla, subphyla, and major classes of marine animals. (No generally accepted numbers for the total species of reef animals by phyla are currently known.)

Marine Animals Ranked by Diversity

Protozoans (80,000 species, incl. freshwater)
Mollusks (50,000 species)
Crustaceans (39,000 species)
Flatworms (18,500 species)
Fishes (14,600 species, incl. sharks & rays)
Segmented Worms (12,000 species)
Cnidarians (9,000 species)
Echinoderms (6,000 species)
Sponges (5,000 species)
Bryozoans (5,000 species)

Source: Ruppert & Barnes, 1994; Nelson, 1994. Some invertebrate totals include freshwater or terrestrial species.

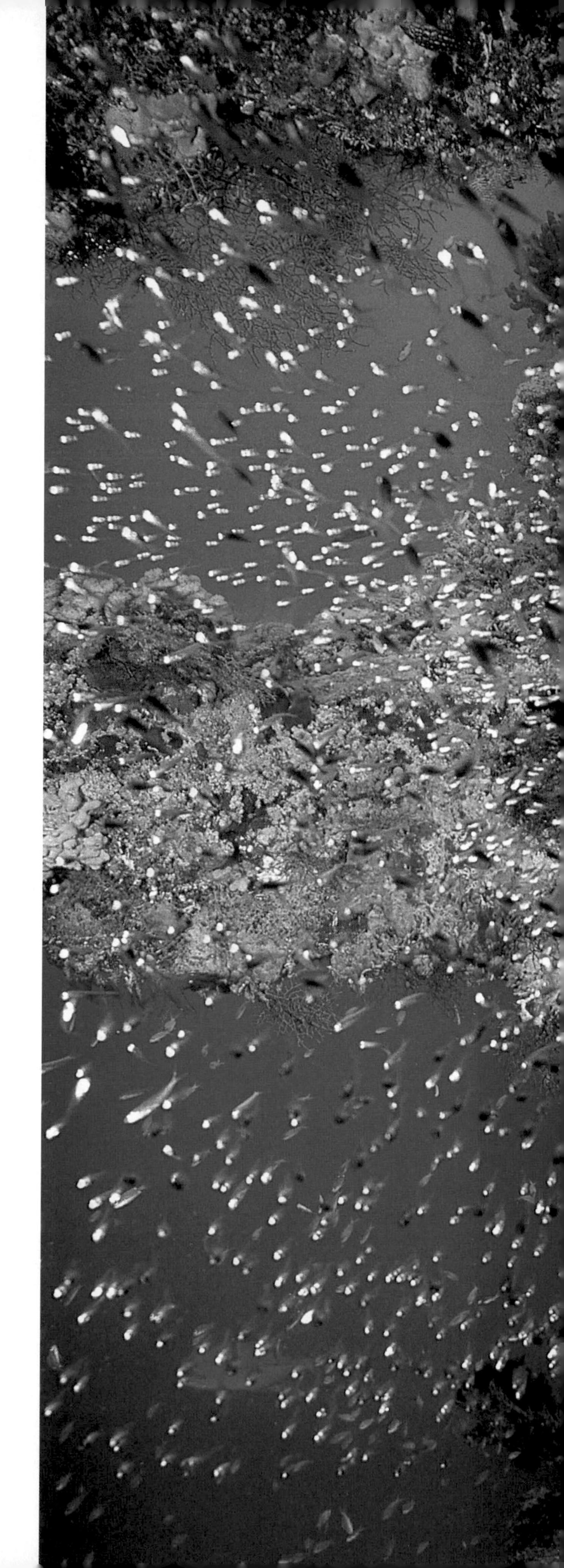

Still counting: the profusion of life on an Indo-Pacific reef illustrates the challenge facing biogeographers, who have barely begun to identify all the species on the world's reefs. More than 90% of all Indo-Pacific reefs are uncataloged.

WEBS OF BIODIVERSITY: PLANTS & ANIMALS OF THE REEF

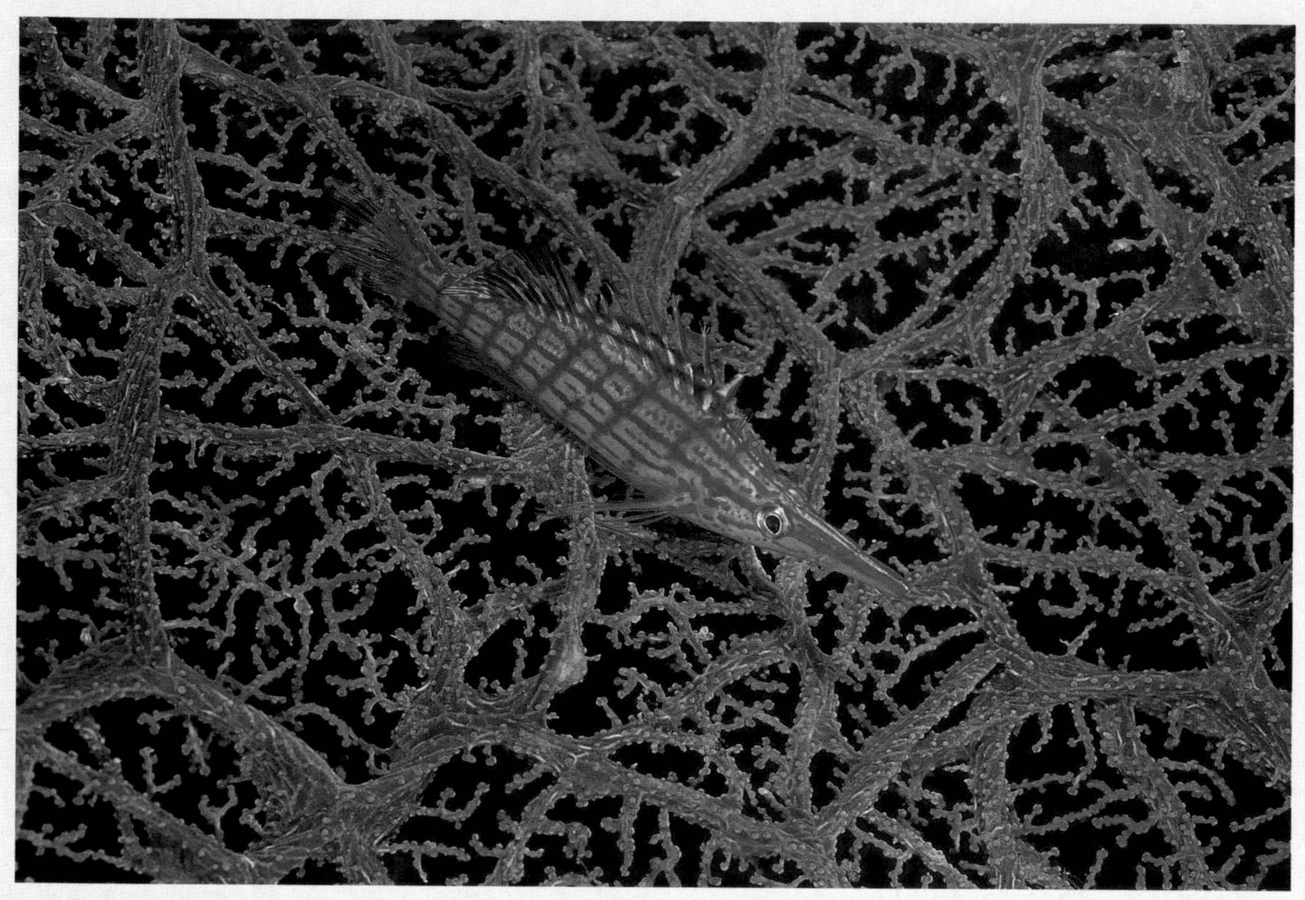

Longnose Hawkfish (*Oxycirrhites typus*) rests on the branches of a colorful sea fan, or gorgonian.

Coral reefs are often ranked with tropical rainforests as Earth's pinnacles of biological complexity and species diversity. This is the result of millions of years of evolution during which species survival was dependent on effective adaptations made in the face of major geologic and climatic changes and fluctuating sea levels.

Evolutionary biologists continue to debate how we came to have so many species, but whatever the final explanation—if one is ever fully agreed upon—those of us who try to be more perceptive observers of reef life must have some comfort with the system of names used by all marine biologists.

Taxonomy is nothing more than the system used by our one species—*Homo sapiens*—to sort all other life forms into similar groupings, or **taxa**, starting with the broadest—Kingdoms—and ending up with the most specific—species. This is a binomial system, in which each valid species is assigned a two-name identifier that states its genus and species. The Longnose Hawkfish above, for example, is *Oxycirrhites typus.* Here is how it fits into the grand scheme of classification in the Animal Kingdom:

Phylum Chordata (animals with notochords)
Class Actinopterygii (all bony fishes)
Order Perciformes (most reef fishes)
Family Cirrhitidae (hawkfishes)
Genus *Oxycirrhites* (closely related hawkfishes)
Species *typus* (the Longnose Hawkfish)

Note that proper usage calls for the genus to be capitalized and the species to begin with a lower case letter. Both genus and species are italicized.

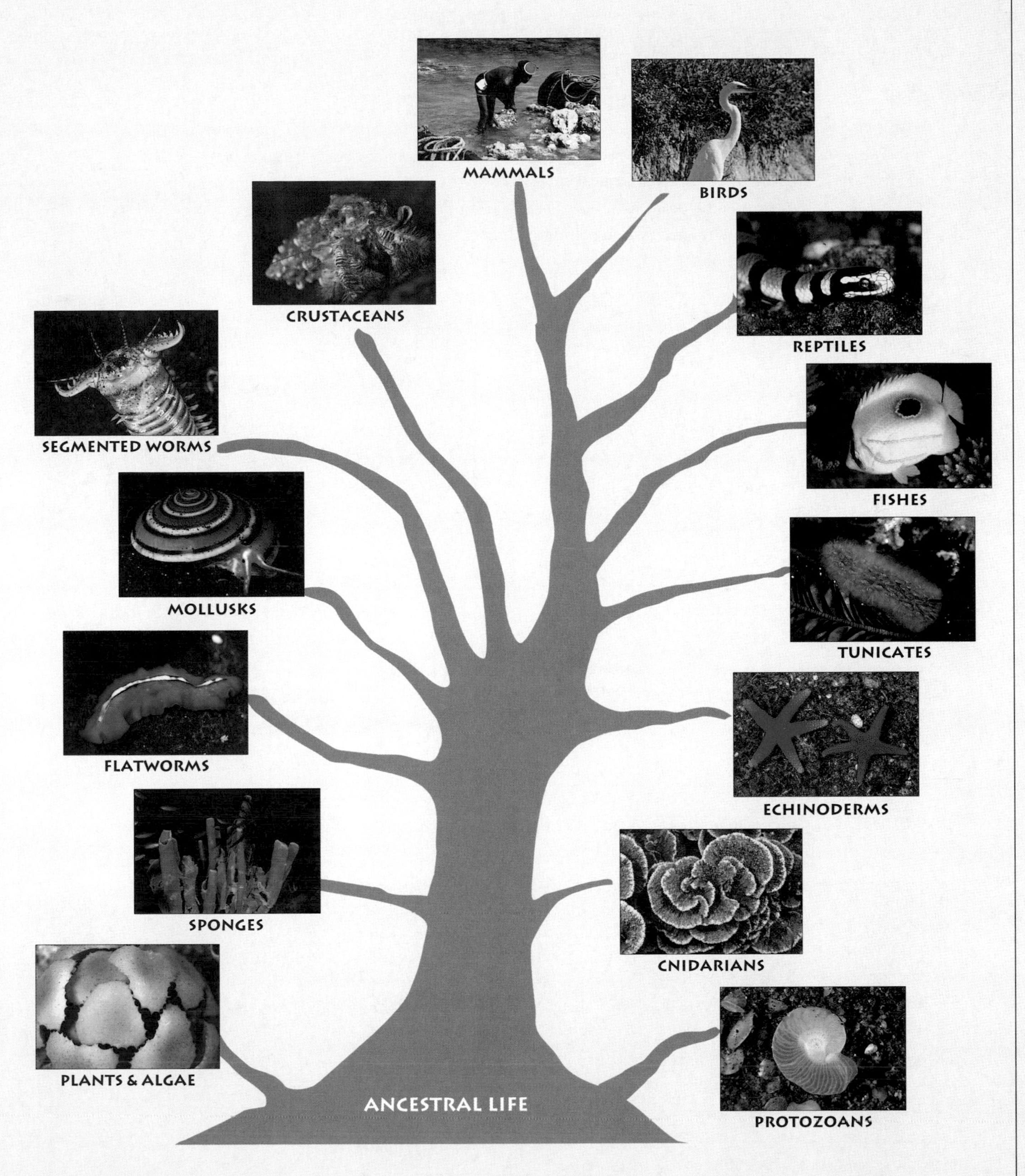

LIFE ON CORAL REEFS: A SIMPLIFIED EVOLUTIONARY TREE

(after Allen & Steene, 1994)

PLANTS

DIVISION MAGNOLIOPHYTA

FLOWERING PLANTS

***NUMBER OF LIVING SPECIES:** 230,000 (a fraction of a percent are marine).

COMMON CHARACTERISTICS OF SEAGRASSES & MANGROVES: marine plants with the same basic structure as terrestrial, flowering plants; adapted to saline water; able to grow when submerged; anchoring system to withstand wave action and tidal currents; water-pollinated.

SEAGRASSES (HELOBIAE)
Halophila sp.

MANGROVES (RHIZOPHORACEAE)
Rhizophora sp.

ALGAE

DIVISION CHLOROPHYTA

GREEN ALGAE

NUMBER OF LIVING SPECIES: 9,000-12,000, about 10 percent marine.

COMMON CHARACTERISTICS: contain chlorophylls *a* and *b*; starch stored inside chloroplast.

CAULERPAS (CAULERPACEAE)
Caulerpa cf. *racemosa*

GREEN ALGAE (CHLOROPHYTA)
Halimeda sp.

***NOTE:** number of species in this section may include freshwater and terrestrial when marine totals are unknown.

GREEN ALGAE (CHLOROPHYTA)
Unidentified green alga

Sea grapes *Caulerpa racemosa* with
Calcareous green alga *Halimeda* sp.

DIVISION CHROMOPHYTA

DIATOMS, GOLDEN ALGAE, BROWN ALGAE, KELPS

NUMBER OF LIVING SPECIES: 18,000, freshwater and marine.

COMMON CHARACTERISTICS: most contain chlorophyll *a*; carotenoids present; starch stored outside chloroplast.

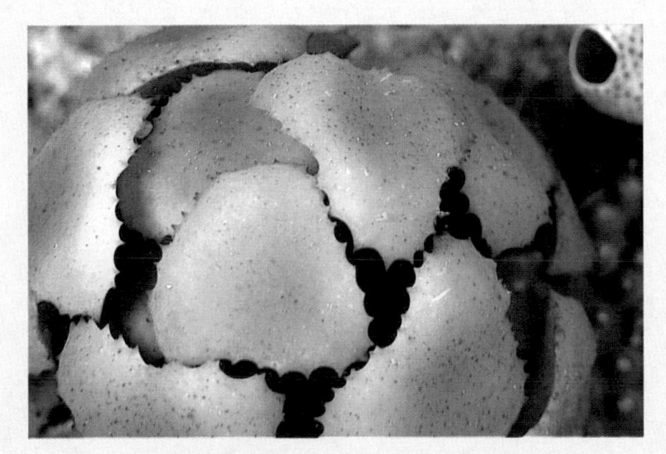

BROWN ALGAE (PHAEOPHYCEAE)
Turbinweed *Turbinaria* sp.

BROWN ALGAE (PHAEOPHYCEAE)
Dictyota sp.

BROWN ALGAE (PHAEOPHYCEAE)
Unidentified brown alga

BROWN ALGAE (PHAEOPHYCEAE)
Dictyota sp.

DIVISION RHODOPHYTA

RED ALGAE OR CORALLINE ALGAE

NUMBER OF LIVING SPECIES: 4,100, predominantly marine.
COMMON CHARACTERISTICS: mostly photosynthetic but about one-third parasitic; starch occurs outside chloroplast; coralline algae contribute to coral reefs and coral sands.

Coralline red alga *Lithophyllum* sp. with
Green alga *Halimeda* sp.

RED ALGAE (RHODOPHYTA)
Crustose red alga *Peyssonnelia* sp.

PROTOZOANS

ORDER FORAMINIFERIDA

FORAMINIFERS

NUMBER OF LIVING SPECIES: perhaps 7,000, many still unidentified.
COMMON CHARACTERISTICS: single-celled organisms, mostly motile, a few sessile; mostly benthic; shell (test) is made of secreted organic material or cemented foreign particles or calcium carbonate. Although single-celled, most are multichambered; chambers are added in a distinct, symmetrical pattern.

FORAMINIFERS (FORAMINIFERIDA)
Archaias sp.

FORAMINIFERS (FORAMINIFERIDA)
cf. Family Langenidae

ANIMALS

PHYLUM PORIFERA

SPONGES

NUMBER OF LIVING SPECIES: 5,000, mostly marine, many still unidentified.

COMMON CHARACTERISTICS: the most primitive of the multi-celled animals; sessile (not free-moving, attached to substrate); most movement of body parts too slow to be visible to naked eye; simple body plan with a system of internal water canals; body supported by spicules (small needlelike or rodlike structures embedded in the tissue).

NOTEWORTHY BEHAVIORS: remarkable water-pumping abilities (one study showed a small sponge circulating 2,250 times its own body volume of water each day); feed on extracted microscopic particulate matter, including organic material, bacteria, various types of fine plankton; some marine sponges harbor symbiotic, photosynthetic zooxanthellae or cyanobacteria.

SILICEOUS SPONGES (DEMOSPONGIAE)
Tube sponge cf. *Kallypilidion* sp.

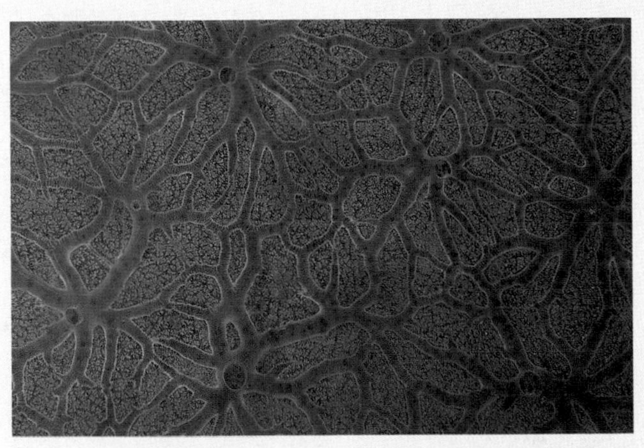

SILACEOUS SPONGES (DEMOSPONGIAE)
Encrusting sponge *Clathria* sp.

SILACEOUS SPONGES (DEMOSPONGIAE)
Red sponge *Monanchora ungiculata*

PHYLUM CNIDARIA

HYDRAS, JELLYFISHES, SEA ANEMONES, CORALS

NUMBER OF LIVING SPECIES: more than 9,000, mostly marine.

COMMON CHARACTERISTICS: radial symmetry; a gutlike gastrovascular cavity (coelenteron) with a common mouth/anus; tentacles surrounding the mouth aid in the capture of food; tentacles armed with stinging structures known as nematocysts; two predominant body forms: the sessile polyp, such as the corals, and the free-swimming medusa, such as the jellyfishes.

NOTEWORTHY BEHAVIORS: all are carnivorous, feeding on plankton of various types and sizes; many have symbiotic zooxanthellae embedded in their tissues that provide an important source of energy; some, especially members of the Classes Hydrozoa, Cubozoa, and Scyphozoa, can deliver painful or even fatal stings to humans; those that secrete calcium carbonate (especially the stony corals) are primary contributors to coral reef formation.

Class Hydrozoa (Hydrozoans, Hydroids, Hydrocorals): 2,700 species

THECATE HYDROIDS (LEPTOMEDUSAE)
Feathery hydroid *Zygophylax* sp.

FIRE CORALS (MILLEPORIDAE)
Fire coral *Millepora* sp.

Class Scyphozoa (Jellyfishes): 200 species

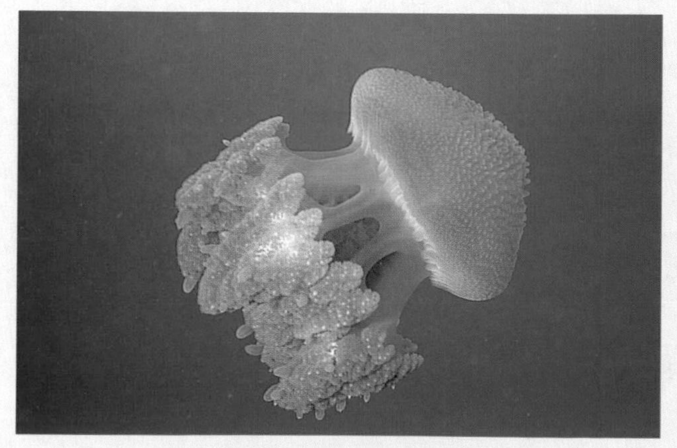

JELLYFISHES (SCYPHOZOA)
Spotted Jellyfish *Phyllorhiza punctata*

JELLYFISHES (SCYPHOZOA)
Upside-down Jellyfish *Cassiopeia andromeda*

Class Anthozoa (Sea Anemones, Stony Corals, Soft Corals, Gorgonians, Sea Pens, Sea Pansies): 6,000 species

SEA ANEMONES (ACTINIARIA)
Unidentified anemone

ZOANTHIDS (ZOANTHIDEA)
Sea Mat or Button Polyps cf. *Zoanthus* or *Palythoa*

STONY CORAL (SCLERACTINIA)
Lettuce coral *Montipora* sp.

SOFT CORALS (NEPHTHEIDAE)
Tree coral *Scleronephthya* sp.

PENNATULACEANS (PENNATULACEA)
Sea pen *Virgularia* sp.

HORNY CORALS (GORGONACEA)
Blue gorgonian *Acalycigorgia* sp.

PHYLUM CTENOPHORA
COMB JELLIES, SEA WALNUTS

NUMBER OF LIVING SPECIES: 50.
COMMON CHARACTERISTICS: delicate, nearly transparent, typically pelagic marine animals that drift in coastal and oceanic waters; bioluminescent.
NOTEWORTHY BEHAVIORS: propel themselves with eight rows of fused cilia or combs (the ctenes); predatory on other forms of plankton.

COMB JELLIES (CTENOPHORA)
Unidentified comb jelly

PHYLUM PLATYHELMINTHES
FLATWORMS

NUMBER OF LIVING SPECIES: 18,500.
COMMON CHARACTERISTICS: soft-bodied, bilaterally symmetrical, dorso-ventrally flattened; head may or may not carry tentacles and eyes; common mouth/anus and no central body cavity.
NOTEWORTHY BEHAVIORS: most move slowly by using tiny hairlike cilia; large species (polyclad flatworms) propel themselves with muscular undulations; some free-living reef species are flamboyantly pigmented as a warning to predators.

Class Turbellaria (free-living flatworms): 3,000 species

POLYCLAD FLATWORM (POLYCLADIDA)
Pseudoceros ferrugineus

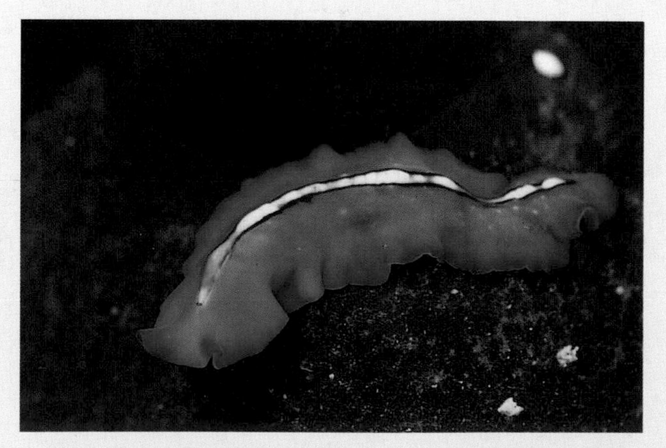

POLYCLAD FLATWORM (POLYCLADIDA)
Pseudoceros bifurcus

PHYLUM NEMERTEA

RIBBON WORMS, PROBOSCIS WORMS

NUMBER OF LIVING SPECIES: 900.

COMMON CHARACTERISTICS: elongated and typically flattened; some appear to be segmented, but these divisions are merely superficial.

NOTEWORTHY BEHAVIORS: some deep-water, pelagic species, but most live in rocky niches, beneath shells and stones, or burrow in soft substrates; move by use of cilia or muscular contraction and expansion; prey on other bottom-dwelling invertebrates.

RIBBON WORMS (NEMERTEA)
Baseodiscus quinquelineatus

RIBBON WORMS (NEMERTEA)
Unidentified nemertean

PHYLUM ECHIURA

SPOON OR TONGUE WORMS

NUMBER OF LIVING SPECIES: 140.

COMMON CHARACTERISTICS: nonsegmented worms with a long extensible proboscis; in the common forms found on reefs, the proboscis splits to form a Y- or T-shaped structure.

NOTEWORTHY BEHAVIORS: some species burrow actively in sand or mud, while others bore into coralline rock; still others occupy empty shells, niches in the coral, or other protective refuges.

TONGUE WORM (ECHIURA)
Unidentified bonellid echiuran

PHYLUM MOLLUSCA

CLAMS, OYSTERS, SQUIDS, OCTOPUSES, SNAILS

NUMBER OF LIVING SPECIES: 50,000.

COMMON CHARACTERISTICS: many secrete a protective dorsal shell of calcium carbonate, although in some groups the shell is much reduced or absent; an anatomical feature in most is the radula, a rasping band covered with rows of teeth; marine mollusks typically have well-developed gills; ventral surface is usually a mucus-secreting foot, a muscular organ used for creeping locomotion; sizes range from minuscule (snails) to the world's largest invertebrates (giant squids).

NOTEWORTHY BEHAVIORS: the generalized marine mollusk is a grazer, moving across the substrate and scraping algae (and other organisms) with its belt-like radula; although many are slow-moving or live attached to substrate, the squids include the fastest-recorded aquatic invertebrates (up to 40 km/hr.).

Class Polyplacophora (Chitons): 800 species

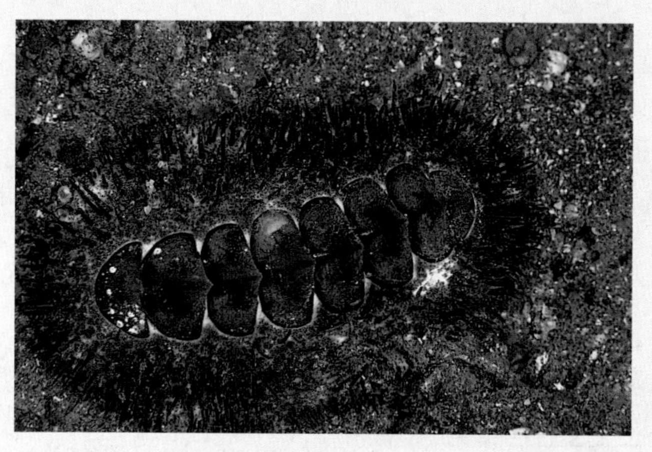

CHITONS (POLYPLACOPHORA)
Acanthopleura spinosa

Class Gastropoda (Snails, Slugs): 30,000 species

SEA SLUGS (OPISTHOBRANCHIA)
Bubble shell *Hydatina physis*

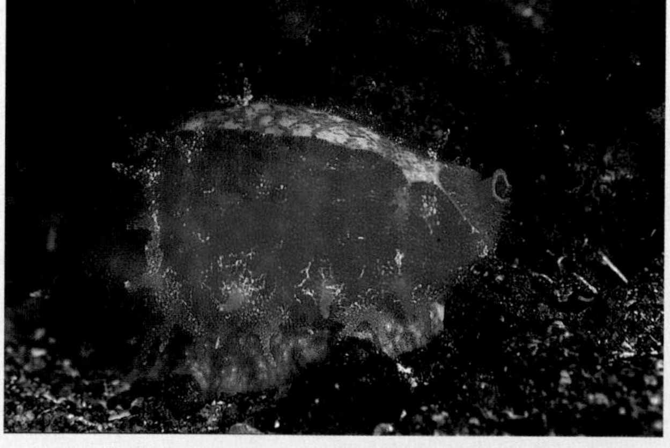

COWRIES (CYPRAEIDAE)
Chinese Cowrie *Cypraea chinensis*

HETEROBRANCHS (HETEROBRANCHIA)
Sundial *Architectonica perspectiva*

TRUE CONCHS (STROMBACEA)
Conch *Strombus thersites*

Class Bivalvia (Clams, Oysters, Mussels): 7,700 species

GIANT CLAMS (TRIDACNIDAE)
Tridacna sp.

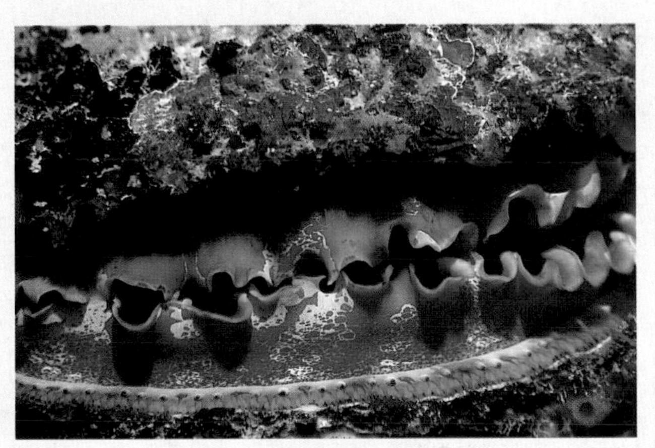

THORNY OYSTERS (OSTREOIDA)
Spondylus varians

Class Cephalopoda (Nautiluses, Cuttlefishes, Squids, Octopuses): 600 species

OCTOPUSES (OCTOPODA)
Octopus luteus

SQUIDS (TEUTHOIDEA)
Reef Squid *Sepioteuthis lessoniana*

PHYLUM ANNELIDA
SEGMENTED WORMS

NUMBER OF LIVING SPECIES: 12,000.

COMMON CHARACTERISTICS: elongate, wormlike animals with segmented bodies; anterior end bears jaws, eyes, tentacles, and antennae; body segments equipped with stiff bristles known as chaetae (used in locomotion and venomous in some polychaetes).

NOTEWORTHY BEHAVIORS: some are raptorial predators with complex jaws, capturing other invertebrates and even fishes; others exist on algae, while many are omnivores, consuming algae, other invertebrates, and detritus.

Class Polychaeta (Polychaete Worms): 8,000 species

LARGE POLYCHAETES (EUNICIDAE)
Bobbit worm *Eunice* sp.

FEATHER DUSTER WORMS (SERPULIDAE)
Christmas Tree Worm *Spirobranchus giganteus*

SPAGHETTI WORMS (TEREBELLIDAE)
Loimiamedusa sp.

PHYLUM ARTHROPODA
ARTHROPODS

NUMBER OF LIVING SPECIES: 750,000.
COMMON CHARACTERISTICS: the most species-rich animal phylum on Earth (includes the insects); perhaps the most successful phylum at adapting to both terrestrial and marine environments; segmented bodies covered with a proteinaceous exoskeleton, strengthened with chitin and calcium carbonate; jointed limbs serve various specialized functions as antennae, legs, pincers, or claws. **Note:** some taxonomists now regard Arthropoda as a superphylum with four distinct phyla: the extinct Trilobites (Trilobita), the chelicerates (Chelicerata), the crustaceans (Crustacea), and the insects, millipedes, and centipedes (Uniramia).
NOTEWORTHY BEHAVIORS: alternating leg movements; sense organs like hairs, bristles, and eyes permit monitoring of environment; dioecious reproduction.

Class Pycnogonida (Sea Spiders): 1,000 species

SEA SPIDERS (PYCNOGONIDA)
cf. *Anoplodactylus* sp.

SUBPHYLUM CRUSTACEA (COPEPODS, BARNACLES, CRABS, SHRIMPS, LOBSTERS)
39,000 species

Class Copepoda (Copepods): 8,400 species

COPEPODS (COPEPODA)
Parasitic copepod attached to fish body

Class Cirripedia (Barnacles): 900 species

PYRGOMATID BARNACLES (CIRRIPEDIA)
Coral barnacles with feeding tentacles

Class Malacostraca (Shrimps, Prawns, Lobsters, Crabs, Amphipods, Isopods): 22,000 species

HINGEBEAK PRAWNS (RHYNCHOCINETIDAE)
Rhynchocinetes durbanensis

SPINY LOBSTERS (PALINURIDAE)
Painted Rock Lobster *Panulirus versicolor*

HERMIT CRABS (ANOMURA)
Aniculus aniculus

AMPHIPODS (AMPHIPODA)
Ladybug amphipods *Cyproidea* sp. on gorgonian

SKELETON SHRIMPS (AMPHIPODA)
Caprellid amphipod (Caprellidae) on gorgonian

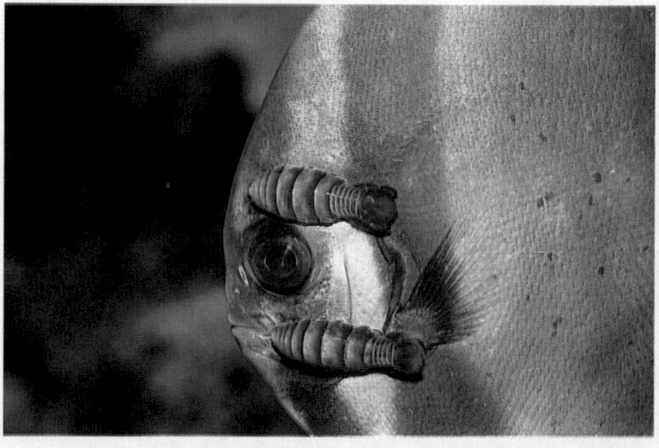

ISOPODS (ISOPODA)
Cymathoid isopods attached to head of batfish

PHYLUM BRYOZOA

BRYOZOANS, MOSS ANIMALS

NUMBER OF LIVING SPECIES: 5,000, mostly marine.

COMMON CHARACTERISTICS: tiny, sessile, colonial animals; form large, variable colonies that may be branching or meshlike.

NOTEWORTHY BEHAVIORS: colonize hard substrates like rock, coral, sand, and mangrove roots; individuals are not interconnected, but typically react to external stimuli simultaneously, retracting into their cases as one.

Bryozoan colony *Triphyllozoon* sp.

PHYLUM ECHINODERMATA

FEATHER STARS, SEA STARS, BRITTLE STARS, SEA URCHINS, SEA CUCUMBERS

NUMBER OF LIVING SPECIES: 6,000.

COMMON CHARACTERISTICS: body has pentamerous radial symmetry (five parts around a central axis); internal skeleton of calcareous ossicles that may be flexible, as in sea stars, or fused into a rigid shell (test), as in sea urchins.

NOTEWORTHY BEHAVIORS: locomotion achieved with the aid of tiny suckerlike pods that are part of a unique vascular system, a complex network of fluid-filled canals and appendages in the body wall.

Class Asteroidea (Sea Stars, Starfish): 1,500 species

SEA STARS (ASTEROIDEA)
Left: *Fromia pacifica* Right: *Fromia* cf. *milleporella*

SEA STARS (ASTEROIDEA)
Crown-of-Thorns Starfish *Acanthaster planci*

Class Ophiuroidea (Basket Stars, Serpent Stars, Brittle Stars): 2,000 species

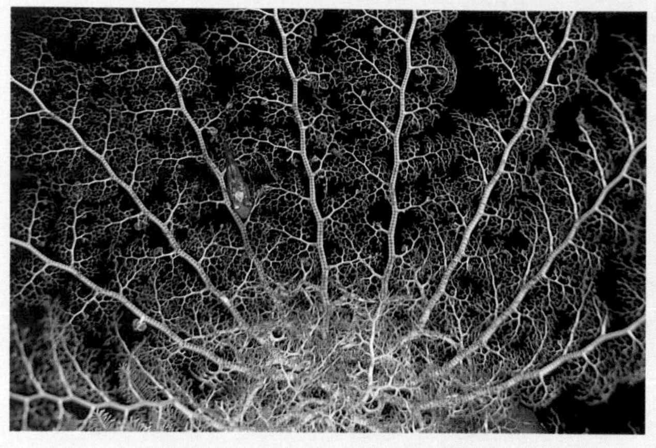

BASKET STARS (OPHIUROIDEA)
Night-feeding basket star *Astroboa nuda*

BRITTLE STARS (OPHIUROIDEA)
Ophiothrix sp.

Class Echinoidea (Sea Urchins, Heart Urchins, Sand Dollars): 950 species

SEA URCHINS (ECHINOIDEA)
Mespilia globulus

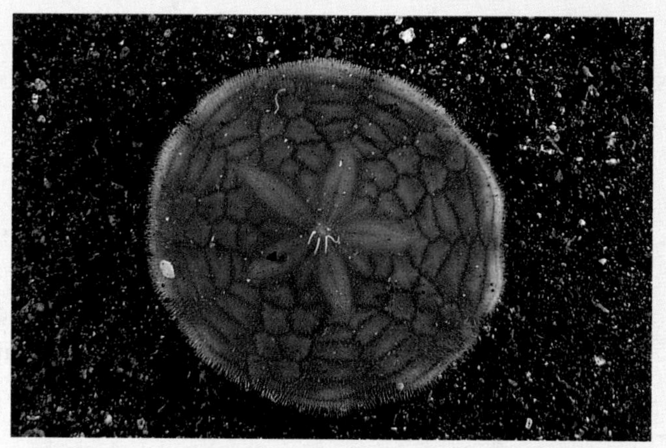

SAND DOLLARS (ECHINOIDEA)
cf. *Peronella* sp.

Class Holothuroidea (Sea Cucumbers): 900 species

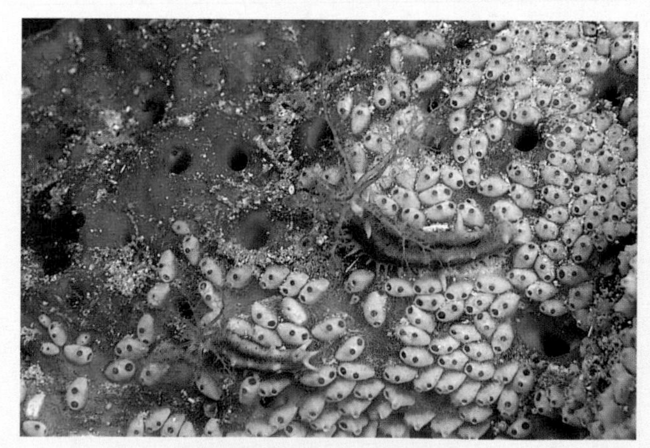

SEA CUCUMBERS (HOLOTHUROIDEA)
Colochirus robustus

SEA CUCUMBERS (HOLOTHUROIDEA)
Stichopus cf. *horrens*

Class Crinoidea (Crinoids, Feather Stars, Sea Lilies): 550 species

FEATHER STARS (CRINOIDEA)
cf. *Comanthina* sp.

FEATHER STARS (CRINOIDEA)
cf. *Comantheria* sp.

PHYLUM CHORDATA
TUNICATES, VERTEBRATES

NUMBER OF LIVING SPECIES: 44,000.
COMMON CHARACTERISTICS: members of the large Subphylum Vertebrata (true vertebrates), have a segmented backbone that protects a spinal cord; two small subphyla, including the tunicates or sea squirts, have a notochord (a hollow dorsal tube analogous to a true spinal cord) during part of their life cycle.
NOTEWORTHY BEHAVIORS: true vertebrates have evolved some of the most complex anatomies, motor skills, intelligences, and behaviors in the Animal Kingdom. They are Earth's apex predators (although some biologists point out that the sheer mass and enormous molecular diversity of bacteria make them the dominant life forms on our planet).

SUBPHYLUM UROCHORDATA (TUNICATES)
1,600 species

Class Ascidiacea (Sea Squirts)

SEA SQUIRTS (ASCIDIACEA)
Didemnidae

Class Thaliacea (Salps, Pyrosomes)

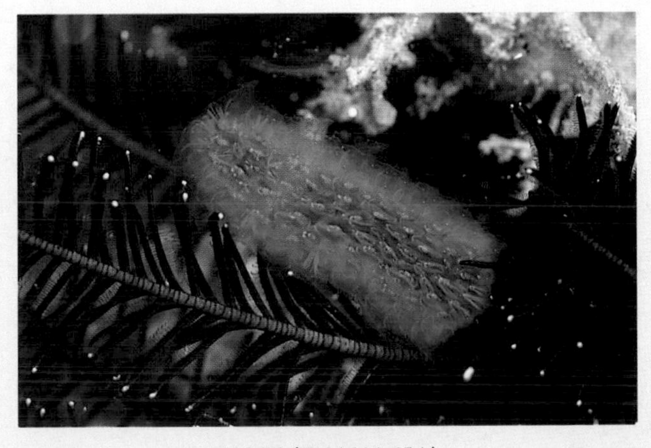

SALPS (THALIACEA)
cf. *Pyrosoma* sp.

SUBPHYLUM VERTEBRATA (VERTEBRATES): 42,000 species

Class Chondrichthyes (Sharks, Rays): 900 species

SHARKS (SUPERORDER SQUALOMORPHII)
Gray Reef Shark *Carcharhinus amblyrhynchos*

RAYS (SUPERORDER RAJOMORPHII)
Manta Ray *Manta birostris*

Class Actinopterygii (formerly Osteichthyes) (Bony Fishes)
23,600 species (about 13,700 marine; 9,900 freshwater)

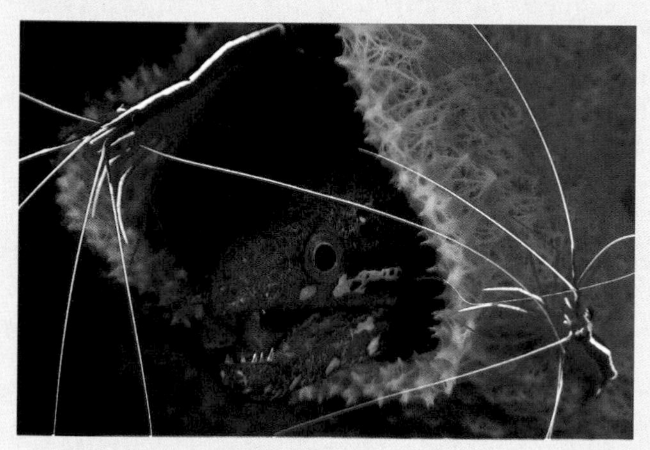

MORAY EELS (MURAENIDAE)
Barredfin Moray *Gymnothorax zonipectus*

SQUIRRELFISHES & SOLDIERFISHES (HOLOCENTRIDAE)
Whitetip Soldierfish *Myripristis vittata*

SEAHORSES & PIPEFISHES (SYNGNATHIDAE)
Common Seahorse *Hippocampus kuda*

SCORPIONFISHES & LIONFISHES (SCORPAENIDAE)
Leaf Scorpionfish *Taenianotus triacanthus*

GROUPERS & ANTHIAS (SERRANIDAE)
Panther Grouper or Barramundi Cod *Cromileptes altivelis*

CARDINALFISHES (APOGONIDAE)
Pajama Cardinalfish *Sphaeramia nematoptera*

SNAPPERS & FUSILIERS (LUTJANIDAE)
Checkered Snapper *Lutjanus decussatus*

BUTTERFLYFISHES (CHAETODONTIDAE)
Bennett's Butterflyfish *Chaetodon bennetti*

ANGELFISHES (POMACANTHIDAE)
Emperor Angelfish *Pomacanthus imperator* (juv.)

DAMSELFISHES (POMACENTRIDAE)
Maroon or Spinecheek Anemonefish *Premnas biaculeatus*

NOTE: fishes shown represent the most-populous coral reef fish families.

WRASSES (LABRIDAE)
Yellowtail Coris *Coris gaimard*

PARROTFISHES (SCARIDAE)
Bicolor Parrotfish *Cetoscarus bicolor* (male)

BLENNIES (BLENNIDAE)
Bigspot Blenny *Crossosalarias macrospilus*

GOBIES (GOBIIDAE)
Twinspot Goby *Signigobius biocellatus*

SURGEONFISHES (ACANTHURIDAE)
Yellowfin Surgeonfish *Acanthurus xanthopterus*

TRIGGERFISHES (BALISTIDAE)
Clown Triggerfish *Balistoides conspicillum* (juv.)

NOTE: fishes shown represent the most-populous coral-reef fish families.

Class Reptilia (Reptiles): 5,000-6,000 species
MARINE EXAMPLES: sea turtles, sea snakes

SEA TURTLES (CHELONIDAE)
Hawksbill Turtle *Eretmochelys imbricata*

SEA SNAKES & KRAITS (LATICAUDIDAE)
Banded Sea Snake *Laticauda colubrina*

Class Avia (Birds): 10,000 species
MARINE EXAMPLES: reef heron, frigate bird, pelican

HERONS & EGRETS (ARDEIDAE)
Pacific Reef Egret *Egretta sacra*

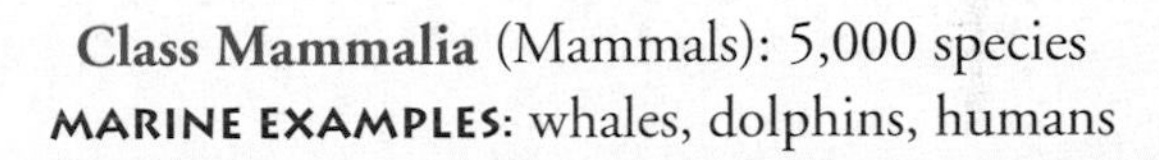

Class Mammalia (Mammals): 5,000 species
MARINE EXAMPLES: whales, dolphins, humans

DOLPHINS (DELPHINIDAE)
Spinner Dolphins *Stenella longirostris*

HUMANS (HOMINIDAE)
Coral collectors on a Maldives reef *Homo sapiens*

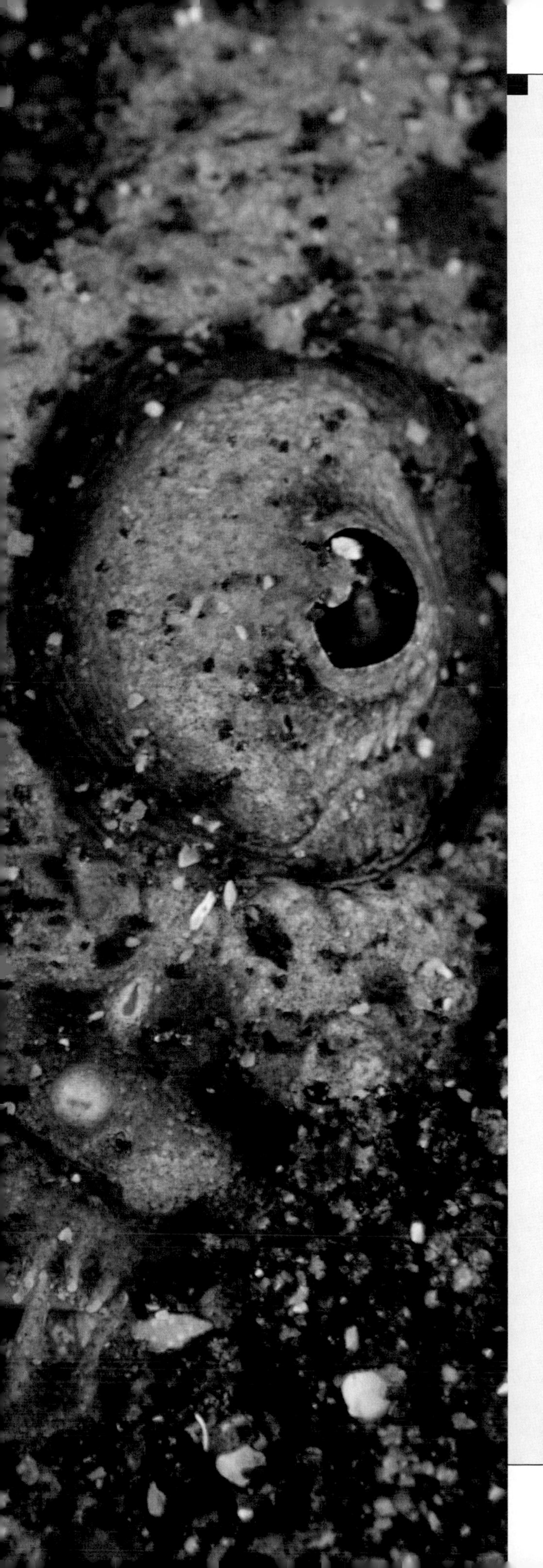

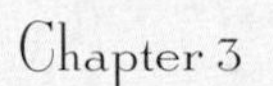

Chapter 3

Near-Reef Habitats

Life in the Sand, Silt, and Mangroves

"He goes on a great voyage that goes to the bottom of the sea."
Proverb

During the summers of my youth, I went clamming with my father on the north shore of Long Island not far from where we lived. I looked forward to these outings, not only because I was fond of steamed clams, but because my father and I shared a love of the sea that my mother never fully understood. We loved smelling the sea air and watching the gulls and trudging around in the mud at low tide looking for the tiny bubbles that gave away a buried clam. The first sign of a bubble would send us digging through the mud with our hands to find the buried treasure. Shouts of "I got one!" resounded along the shore. It was a busy time when the tide went out. Later in the day, when our buckets were full, we headed home to steam our hard-earned booty.

That was a long time ago, but those memories of digging for clams are very special. Back then, clams were all I expected to see in the sand and clams were all I found, save for the occasional crab that scurried along the beach. Since then I've learned many marine secrets, among them that there are more species living in the sand than perhaps anywhere else in the sea.

Sand dweller: a Stargazer (*Uranoscopus* sp.) lies buried to its eyeballs in dark sand, ready for a lunging attack.

Moon Snail (*Natica orientalis*): emerging primarily at night from its hiding place in the sand, this predatory mollusk displays an enlarged head flap that aids its ability to burrow into the soft substrate. Moon Snails prey on other mollusks, including clams and other snails. Many near-reef dwellers have specialized anatomies—for foraging, breathing, and burrowing—to thrive in soft-bottomed habitats.

Fanworm (Polychaeta): using its feathery feeding tentacles to sieve tiny food particles and fine plankton from the water, this soft-bodied marine relative of the common and less flamboyant earthworm is protected in a tube that it secretes. Whether clinging to rock or buried in sand, tubeworms are able to retract their tentacles with blinding speed if a change in light or water pressure suggests the presence of a threat.

Peacock Sole (*Pardachirus pavoninus*): perfectly adapted to a sandy environment, this fish is able to camouflage itself with astonishing changes in skin patterns and pigment intensities. As with other flattened bottom-dwelling soles and flounders, the paired dorsal eyes are a unique adaptation, in which one eye has migrated from the bottom side of the fish to the top. Elevated eyes and nostrils allow the animal to function perfectly while partially buried in sand.

Ambon Scorpionfish (*Pteroidichthys amboinensis*): displays skin filaments that obscure its presence on the open bottom.

THE MONOCHROMATIC PANORAMA OF A SAND or mud bottom can give one the impression of a barren wasteland—an uninteresting environment devoid of all life. Compared to the coral reef itself, the nearby habitat can appear dull and lifeless to the casual observer.

In fact, nothing could be further from the truth. What better place for a marine animal to pass some or most of its life than hidden in the safety of a soft bed of sand or silt, an expanse of thick seagrass, or among the roots of a mangrove swamp—separated from the everyday eat-or-be-eaten competition of life on the reef? Why expose oneself to the fierce predation of the reef or the open water column when the safety of other off-reef habitats beckons?

Non-reef habitats play an important role in the healthy functioning of a coral reef ecosystem. These areas often act as nutrient sinks, catching detritus and wastes from shore and reef alike. The higher nutrient levels breed heavy populations of small organisms that serve as important food sources for many reef juveniles that may pass part of their early life cycle off the reef. Biologists know that where there is abundant food there is abundant life, and some of these habitats are every bit as interesting to explore as the reef itself.

Sandy bottoms occur in lagoons, bays, channels, within reef complexes, and between fringing reefs and beaches. I confess a particular affection for these habitats, where we have found so many unique and unusual marine creatures.

The shifting sediments provide plenty of hiding places. More common, though, are the vast unseen communities of invertebrates with their myriad networks of tunnels and burrows just under the surface. These labyrinthine communities house tens of thousands of species that spend a good part of their lives both underwater and underground. Animals that live in, on, or under the substrate are called **benthic organisms**, as opposed to **pelagic organisms**, those that pass the better part of their lives in the water column. Like flora and fauna elsewhere, benthic organisms have, over time, adapted to their environment with physical

Saccoglossan sea slug (*Elysia ornata*) grazes a patch of macroalgae from which it will extract live chloroplast cells that concentrate in its own tissue, giving it a bright green coloration and possibly serving as a source of nutrition.

Blackfinned Snake Eel (*Ophichthus melanochir*) lies in wait for passing prey fishes or crustaceans, hiding its nearly 1 m (3.3 ft.) length in the dark sand substrate. This eel roams freely at night, when large predators are less of a threat.

Black Shrimp Goby (*Cryptocentrus fasciatus*) and its partner for life, a commensal alpheid shrimp, share a burrow in the sand. The nearly blind shrimp maintains the tunnel, while the goby is the hunter and sentinel.

Oriental Helmet Gurnard (*Dactyloptena orientalis*), a common Indo-Pacific species on sandy expanses near reefs, hunts the bottom for invertebrates by "walking" on its modified pectoral and ventral fins.

modifications and specialized techniques that aid in their struggle for survival.

Soft-bottom communities teem with life—but much of it goes unseen by the untrained or unaided eye. Much of this life is microscopic; a magnifying lens or microscope is essential to appreciate fully the biological diversity of sand and silt. There are thousands of bottom-dwelling animals that are extremely small, with body diameters of about one-tenth of a millimeter.

Interstitial animals include protozoans, hydroids, tiny crustaceans, mollusks, countless worms, and other wormlike animals. These animals, also known as **meiofauna**, spend their entire lives in the water that fills the spaces between the grains of sand. Some have adapted to the constraints of their environment by developing bodies that are long and thin; others are hard-shelled, or armored, to protect their soft inner tissues from the abrasive effects of the substrate and to fend off predatory meiofauna.

Mangroves are tropical to semitropical bushes or trees from several plant families that were once estimated to line up to 70 percent of tropical shorelines. These dense forests and thickets are uniquely able to survive in a saline (salty) environment, and their **prop roots** arching down into the substrate provide a thicket of underwater branches. Here a huge amount of organic matter accumulates and many organisms congregate in relative safety from large predators.

The organic matter in a mangrove swamp comes from a variety of waste products, dead organisms, runoff from the land, and other terrestrial organic debris, like fallen leaves and branches. It may also be detritus washed out of the reef and pushed inshore by tides and storms.

Estuaries are coastal inlets where freshwater streams or rivers meet the ocean. They typically host unique assemblages of organisms—some expatriates from the reef, others washed down from pure freshwater habitats, and still others permanently suited to life in **brackish waters** with a salt content from about 15 percent to 50 percent of full oceanic salinity. Coral reefs generally form some distance from estuaries, which carry seasonal floodwaters out to sea. These events can cause the demise of corals, which are unable to tolerate the sudden drop in salinity and the deposition of sediments.

Lagoons are closely affiliated with coral reefs, and they support their own unique communities of plants and animals. In the Caribbean, lagoons are often relatively shallow, narrow bodies of protected water, often less than 500 m (1,640 ft.) across. They are often covered with pastures of turtle grass or manatee grass, and are vitally important to many grazing animals.

In the Indo-Pacific region, lagoons can be immense—some easily stretch 40 km (24 mi.) across, with depths to 90 m (295 ft.). The lagoon bottom may have vast expanses of sand composed of crushed coral, foraminifera shells, and the calcareous skeletons of *Halimeda* macroalgae. It may have rubble-covered zones and areas of mud or silt.

Patch reefs and **coral heads** often dot the lagoon, providing microhabitats where fishes and invertebrates cluster. A coral head may be a dense cluster of stony corals or a single large colony crowded with associated organisms at its base and between its branches.

Patch reefs may be significantly larger, with a diverse collection of stony corals and other animals. Like a safe haven in a wide-open wilderness patrolled by large carnivores, both patch reefs and coral heads may boast a tremendous diversity of fishes. One study of a Pacific atoll, for example, found 43 different species of fishes on a single patch reef in the lagoon.

The Florida Reef Tract running along the Atlantic coast of south Florida to the Florida Keys is comprised of more than 6,000 patch reefs. (Some refer to this as a barrier reef, but most reef biologists think of it as a **bank reef system** of relatively deep growth on submerged platforms of ancient limestone.)

Pinnacles are common in some Indo-Pacific lagoons and appear as steep knolls that are often topped with stony corals.

Seagrass beds, as mentioned earlier, are quiet meadows of green aquatic grasses, soft-bodied species of macroalgae, and calcareous algae. Keen-eyed snorkelers and divers can find myriad corals, invertebrates, and fishes, including many juvenile reef species, here.

Marine aquarists are often astonished to learn that a high proportion of "reef" livestock they see actually comes from shallow non-reef areas. With calmer waters and easier collecting conditions, these off-reef locations provide prime picking grounds for indigenous peoples who make their living by harvesting fishes, small corals and other invertebrates. Many edible species—food fishes, mollusks, echinoderms—are also sought out in lagoons, bays, and, other shallow areas.

Mangrove swamp: fringing tropical shorelines worldwide, the many families and species of mangrove trees are able to dominate a salty environment in which few other terrestrial plants can survive. In addition to stopping erosion, mangrove stands and swamps provide a nutrient sink, trapping leaf litter and detritus in the profusion of prop roots that anchor the trees in the muddy bottom. Many juvenile fishes and invertebrates are swept into these habitats, where large predators seldom penetrate. The clearing of mangroves in the Tropics has severely impacted many coastal ecosystems and reefs.

A protective nursery: for many young fishes and invertebrates, the roots of a mangrove swamp offer a nutrient-rich environment where large, roving predators and schools of voracious fishes are physically excluded.

Sargassum weed: floating in huge, buoyant rafts, the weed can serve as temporary or even permanent habitat for many organisms, such as this Sargassum Frogfish (*Histrio histrio*) (**1**), blending almost perfectly with its surroundings.

Cryptic appearance is a relatively common modification of species living in the seagrass, such as this pipefish (*Corythoichthys* sp.) whose green, flattened body closely resembles a blade of vegetation. Note head and eyes (**1**).

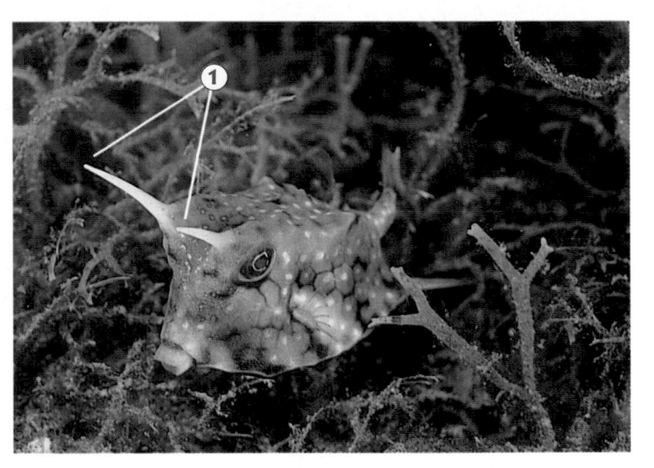

Algae patches tucked on and around reefs offer grazing as well as habitat for a profusion of animals, such as this juvenile Longhorn Cowfish (*Lactoria cornuta*). Note characteristic "horns" (**1**) that suggest its common name.

Seagrass beds harbor many juvenile fishes and others well-adapted to their surroundings. This Seagrass Filefish (*Acreichthys tomentosus*) has assumed the colors of the grasses, disguising its presence both from larger predatory fishes and from the smaller items that serve as its prey.

SOFT-BOTTOM SETTLERS

Soft bottoms are not the dull, lifeless vistas they appear to be at first. In fact they harbor some of the most fascinating creatures in the sea. The fauna presented in this chapter are but a few of the many interesting animals that inhabit the near-shore soft bottoms. There are many more. Although different substrates foster different species, there are many similarities in the way benthic fauna have adapted to their environment.

An awareness of some of the adaptations in a seemingly barren environment makes further faunal observations easier and more interesting.

Stalked eyes, for example, allow many animals to take refuge in the sand or mud, while literally keeping an eye out for potential predators or prey. It takes keen perception, but these telltale eyes can sometimes reveal the presence of a buried flatfish or crustacean.

Similarly, any physical object that breaks the monotony of an open bottom of soft substrate is likely to have attracted one or more animals using it for a foothold or covering shelter.

A single coral head will always reveal a number of hangers on, typically clustering around its base or in its branches. Large anemones anchored in the sand are gathering spots for various fishes, shrimps, crabs, and other affiliated fauna.

Patches of algae or seagrass are always likely locations for hiding fishes and invertebrates, either juveniles that use these areas as a stopping off point on the way to adulthood, or specialized organisms—often perfectly camouflaged—that fit into the beds of green blades of grass or filamentous algae.

Harlequin Crab (*Lissocarcinus laevis*) dwells under the protective umbrella of a Corkscrew Tentacle Anemone (*Macrodactyla doreensis*), whose stinging abilities are well-known to roving fishes that typically give it a wide berth. This anemone anchors its fleshy column deeply in soft sandy or muddy substrates.

Patch reefs or small aggregations of coral heads in an otherwise open lagoon provide crucial shelter to many species, such as this large Blackspotted Moray Eel (*Gymnothorax insigteena*). Stony corals need a firm substrate and cannot proliferate here; those that settle on sandy bottoms may be tumbled over and destroyed in heavy storms.

Reef fringes offer these schooling Bluelined Snappers (*Lutjanus kasmira*) prime hunting grounds for benthic invertebrates and smaller fishes. Too large to seek cover in many of the tight confines of the reef, such fishes form overwhelming packs that confer protection from large piscivores such as sharks and barracudas. Snappers are an important food-fish resource in many reef areas.

Rubble zones are composed of pieces of dead stony coral that are created in tremendous quantities during violent weather. These pieces are interspersed with empty shells and other reef materials. With most of its body tucked into a mucus burrow in the rubble, this mantis shrimp (*Lysiosquilla* sp.) is poised to spear target items that venture within striking distance of its lethal claws(**1**).

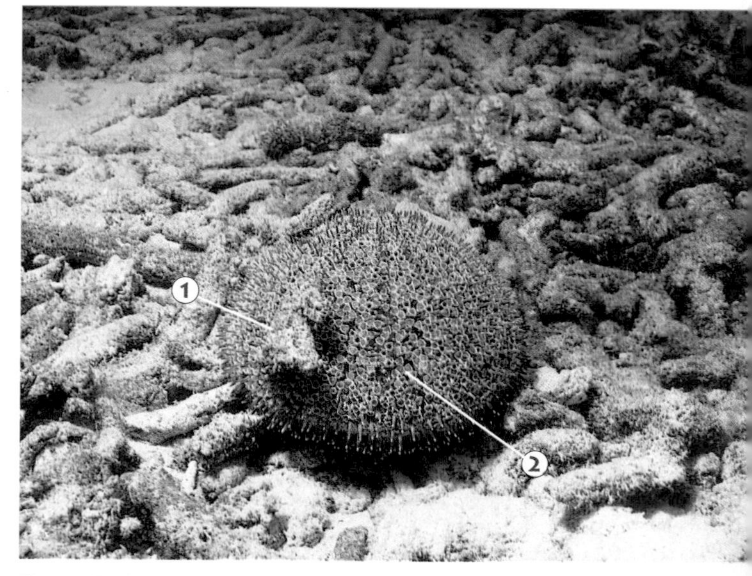

Toxic Flower Urchin (*Toxopneustes pileolus*) is a denizen of rubble zones and sandy bottoms, where it often disguises itself with pieces of dead coral and other debris (**1**). This species has short spines, but they are interspersed with numerous and highly venomous pedicellaria (**2**) that readily sting unwary intruders. Human deaths have been attributed to this urchin.

The pelagic zone is simply the deep blue ocean, lying seaward of the reef. It is patrolled by sea mammals, large cephalopods, and fast-swimming open-ocean fishes, such as this school of silvery Bigeye Trevally (*Caranx sexfasciatus*).

DEEP BLUE SEA

Larger pelagic or open-water fishes and marine mammals can move easily from the oceanic waters that lie seaward of the fore-reef face, and this unsheltered, hostile environment is both a blessing and threat to creatures that spend their adult lives on the coral reef.

The open ocean is crucial to the well-being of the reef, bringing to it a constant influx of waves and currents. These provide essential oxygenation—the churning of surface waters saturates incoming water with oxygen—and prevent stagnation that can quickly kill many delicate reef organisms.

Many corals, for example, appear to be dependent on the to-and-fro motions created by waves and tides to flush away metabolic wastes and excess mucus that they exude.

The force of oceanic water flushing through the nooks and crannies of the reef serves many other purposes by distributing plankton, suspended particulate matter, and dissolved organic wastes to filter feeders and planktivores. Reproducing fishes and invertebrates on the reef very often shed buoyant gametes and/or larvae that are carried out to sea in masses of plankton. These organisms may live and feed in the plankton for weeks, sometimes returning to their "home reef," sometimes drifting great distances to far shores. Ocean currents facilitate the sharing of genetic diversity by connecting these farflung reef systems.

Finally, the deep ocean floor is a sink for organic wastes that may be washed to sea from the reef and other inshore areas. Too much waste accumulating inshore tends to cause **eutrophication**, or an accelerated level of plant or algal growth.

Upwellings periodically bring nutrients from the ocean floor to the surface waters, where they are washed inshore and may spur an explosion of plankton. Upwellings can, however, have deleterious effects, bringing cold water that can shock or kill corals and sometimes precipitating eutrophication events by over-delivering nutrients to the surface. ■

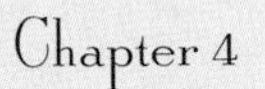

Chapter 4

Living Together

Symbiosis: A Host of Coordinated Reef-Survival Strategies

"Most real relationships are involuntary."
Iris Murdoch,
The Sea, The Sea (1978)

One day while diving in Lembeh Strait in North Sulawesi, I came upon a group of brightly colored sea urchins. One of the urchins had a juvenile Red Emperor Snapper swimming energetically among its spines as juveniles do. While I watched, the young fish plucked bits of food from the surrounding sand and darted back among the urchin's spines when it sensed danger. Returning for a night dive, I found the young fish nestled at the base of the spines while the urchin combed the bottom in search of food.

Juvenile Red Emperors have no defenses of their own. They're too colorful to go unnoticed, but small and agile enough to swim between sea urchin spines. Living within a group of urchins makes it easy for the snapper to feed with little risk.

I continued to observe this pair of unlikely allies over time, and one day I got another surprise. The Red Emperor had just about outgrown its urchin and had

Safe haven: tiny and vulnerable, a juvenile Red Emperor Snapper (*Lutjanus sebae*) lives among the venom-tipped spines of a sea urchin (*Astropyga radiata*), which in turn receives nourishment from the fish's nitrogen-rich wastes.

taken on an overall pinkish tone. The group of urchins had doubled in number and several other adolescent Red Emperors had now joined the pack. As they approach adulthood, young Emperors are believed to migrate to deeper waters in the safety of a group. I missed their actual departure, but can still imagine a mini-migration of these little snappers timidly leaving their urchins behind.

We know why humans live in a society, but what about reef creatures? Although such questions may make a professional biologist wince, they are typical for many of us when we first contemplate the bewildering concentration of life forms on a coral reef.

Do they simply live in close proximity to one another, pursuing their lives independently, or do they form voluntary relationships with each other? Are those relationships interspecific, occurring between different species, or are they confined to their own species?

How does a reef creature go about striking up a relationship with another organism? Are reef creatures trustworthy partners? Does every species have to interact with another? Are there any true hermits in the sea, animals that live completely solitary lives except for feeding on other species? And what do the answers to these questions say about life on the reef? Can we even answer all these questions?

Here we will return to the biologist and the salvation of the scientific approach to mass confusion: first split things—in this case reef relationships—into similar groups, hoping that out of chaos will come understanding.

Symbiosis is the interrelationship between two organisms. It is a catch-all label for the relationship between two or more different organisms that live together. Unfortunately, many us have a faulty recollection of the true definition, and it is a term subject to frequent misuse in the media. A symbiotic relationship is not necessarily beneficial for both organisms involved. In fact, from the point of view of a **symbiont** (a participant in a symbiotic relationship), it can be good, bad, or unimportant. Obviously, more specific terms are needed.

Mutualism is a symbiotic relationship in which both partners derive some benefit from the arrangement. The classic reef example of mutualism is an anemonefish (clownfish) that lives within the tentacles of a sea anemone. Anemonefishes are poor swimmers, slow, and generally unable to defend themselves from predators out in open water. Sea anemones possess microscopic harpoonlike stinging organelles called nematocysts in their mass of tentacles. Most reef fishes know enough to avoid these animals, but those that bumble into the arms of an anemone end up stung to death and eaten—or sent away with unforgettably fiery stinging wounds. In a manner that scientists are still trying to understand, the anemonefish is able to nestle into the tentacles of the sea anemone unscathed and to live out of reach of virtually all predators (see photographs, page 104 and 105).

This relationship works both ways. The fish darts out to capture food, and often inadvertently shares at least part of its catch with the host. The anemone, having very unrefined tastes, is also able to gain nourishment by absorbing the anemonefish's bodily wastes. Additionally, the anemonefish will aggressively defend its host if one of the predatory butterflyfishes that prey on anemones should approach.

Commensalism is a symbiotic relationship in which one partner derives benefit, while the other is neither helped nor harmed. An example of well-known commensal hosts are the sea stars, or starfish. These appear to be solitary species, with individuals going about their business of finding mollusks to consume with little regard for anything else. Closer observation of many sea stars will dispel this misconception. It is not uncommon to turn over a sea star and find several symbionts living there—some are commensal shrimps or crabs, others are parasitic mollusks or scale worms. These symbionts often take on the color of their host, making them difficult for a predator or casual observer to find (see photographs, pages 96 and 101). The small commensals gain a protective shelter while doing nothing to harm their much larger host.

Parasitism is a symbiotic relationship in which one species attaches to another and derives benefit at the expense of its host. An obvious example seen sooner or later by most snorkelers, divers, and marine aquarists is the parasitic fish louse, or parasitic isopod. It is a crustacean that burrows its sucking mouth parts into the flesh of fishes, which are powerless to escape its en-

Mixed schools: single-species shoals or schools are a common sight, but as this reef scene from the Maldives illustrates, different species may join together to feed or avoid predation, for example, squirrelfish (**1**), soldierfish (**2**), anthias (**3**).

ergy-sapping presence (see photograph, page 109).

These are just a few of the tantalizing—or sometimes repulsive—symbiotic arrangements seen among coral reef species. It is important to understand that these relationships are not friendships or feuds, but rather a matter of life and death. Interaction means survival for many marine animals, and these relationships are old beyond our comprehension. It is estimated, for example, that the association of anemones and anemonefishes may date back some 25 million years.

Species interact for many different reasons. Reef creatures are driven by the primal desires to feed, avoid danger, and reproduce. Evolutionary refinements over millennia have seen to it that most species are well developed to do all of these things—albeit in very different ways. So why would a reef animal associate with another reef inhabitant, who may be competing for the same resources of space and food? The first and most obvious answer is because there is something to be gained by the association. Some animals use other species to obtain easier access to food or better shelter than they themselves could obtain. Sometimes the advantage is faster transport or a better location for nutrition, respiration, or reproductive purposes. These are all valid reasons for the interspecific association of reef organisms, however, they are not the only explanations and rarely are these stories as simple as they appear at first glance.

Certainly every species requires another on which to prey, but aside from the predator-prey relationship, there are many species, such as snake eels and octopuses, that are fairly independent throughout their lives, except at breeding times. It is the mutual dependence—and tangled interdependence—of different species that makes life on the reef so interesting. Putting together all the pieces of this never-ending natural history puzzle is a vexing challenge at times—and one that often poses more questions than it answers. ■

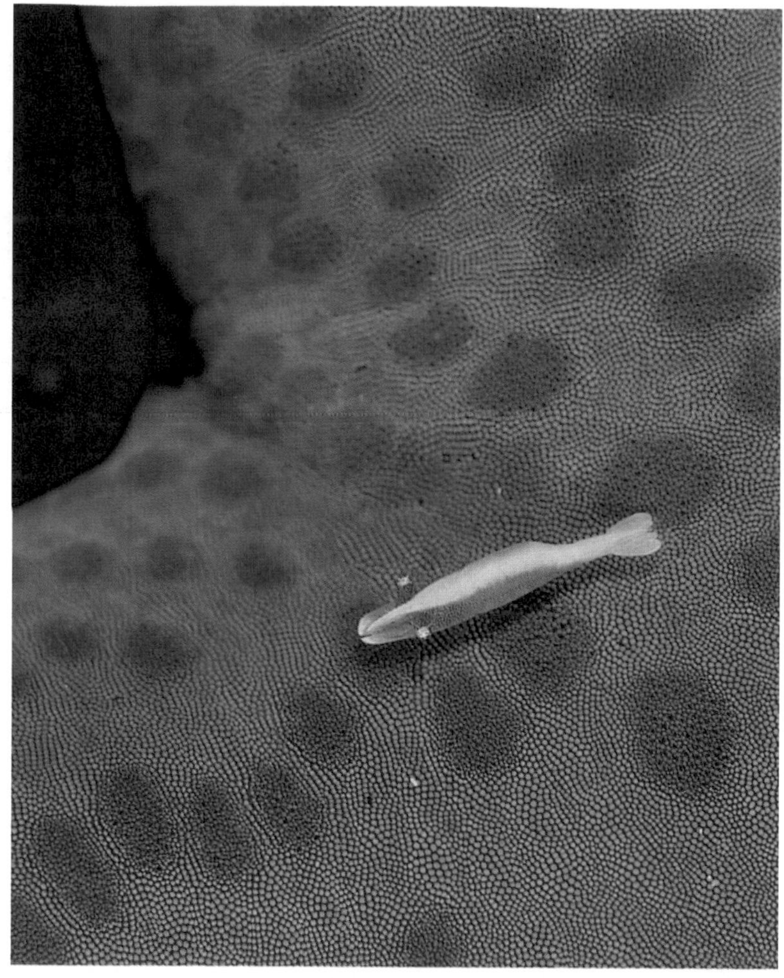

Blue star & symbiont: a Commensal Starfish Shrimp (*Periclimenes soror*) is found on the oral side (bottom) of a Blue Sea Star (*Linckia laevigata*).

TINY COMMENSALS

A great many commensal relationships are not obvious because of the small size of the symbionts. Gorgonians (horny corals) and feather stars (crinoids) host many small crustaceans, mollusks, worms, echinoderms, and fishes in a variety of relationships. In practice, these relationships may be easier to observe than the more obvious ones, like those of sharks and remoras, because of the stationary nature of the host organism.

These hosts live successfully with several symbionts at any given time. Gorgonians and feather stars succeed because they have the capacity to support multiple tiny symbionts by providing them with endless shelter possibilities among their intricate physical structures. It is not unusual to find a pair of squat lobsters, a pair of pontoniine shrimps, a brittle star, and several clingfish, all living in harmony on the same feather star.

Sea stars, also known as asteroids or starfish, frequently are found to have diminutive commensals clinging to their rough, often spiky skins. (These hosts are echinoderms, from the Latin term for "spiny-skinned.")

Among the commensals that commonly land on the sea stars are small shrimps, crabs, worms, and even fishes. Small cardinalfishes have been found hiding among the spines of the venomous Crown-of-Thorns Starfish (*Acanthaster planci*).

As seen in several of the accompanying photographs, commensals often match or camouflage themselves with the colors of their sea star hosts.

Benefits, if any, to the hosts are not known, but the little hitchhikers are able to use their hosts for shelter, transport, and food-finding: when the sea star moves, it may bring the commensals into contact with food sources, and when it feeds, the commensals will greedily share the scraps.

Free ride, no toll: this Commensal Emperor Shrimp (*Periclimenes imperator*) finds safe domicile on the skin of a sea cucumber (*Bohadschia argus*), one of several hosts, including nudibranchs and sea stars, used by this species. The shrimp feeds on detritus and wastes generated by its much larger carrier, which is likely oblivious to its presence.

Vivid obscurity: a deep red Snapping Shrimp (*Synalpheus stimpsonii*) somehow effectively hides in the equally colorful tangle of arms on a filter-feeding feather star (Crinoidea). Small crustaceans are prime targets of many reef carnivores, and their survival may depend on finding effective cover where food items are within reach.

Obscure life: hiding is the only real option for this commensal shrimp (*Periclimenes* sp.), one of more than 240 known species of pontoniine shrimps that are always found living on a host, in this case a feather star (Crinoidea). They are known to feed on smaller commensals that are also attracted to the crinoid, and may nibble at any ectoparasites that attack the host.

Bio-impressionism: a commensal shrimp (*Periclimenes imperator*) displays a modified color pattern that helps it blend into the pigmentation of its host sea cucumber (*Stichopus* sp.). This species has pincers, or chelipeds, that are tipped in purple, but otherwise can assume different color patterns to match the external appearance of the surface where it resides.

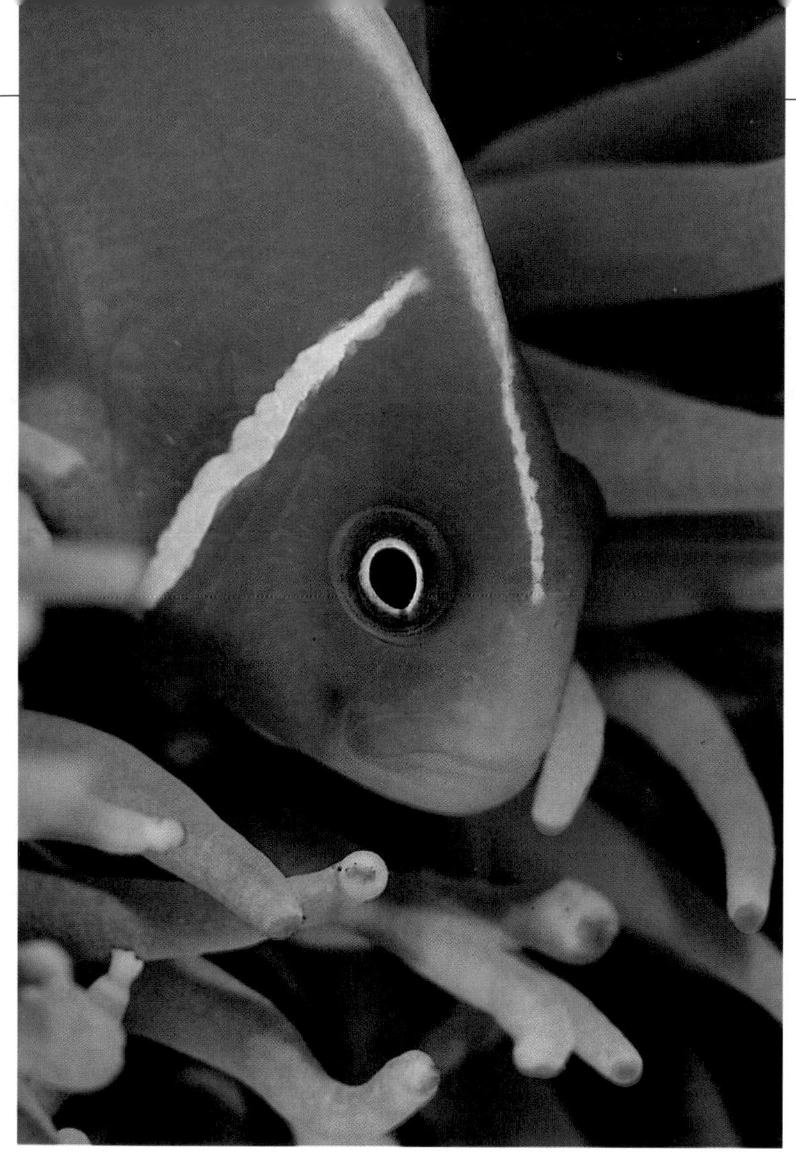

Perfect partners: A Pink Skunk Anemonefish (*Amphiprion perideraion*) seldom ventures far from the venom-tipped tentacles of its host Leathery Sea Anemone (*Heteractis crispa*). This is a classic example of mutualistic symbiosis (mutualism) between two very different animals.

CLASSIC MUTUALISM

In the wild, the anemonefishes simply cannot survive for long in open water. Their swimming style has been described as fluttering or waddling—endearing to human watchers but an invitation to predators that make quick work of any anemonefish that is caught away from a refuge. In the wild, anemonefishes are virtually always found living in close association with a sea anemone, snuggling into its Medusa-like tentacles that make a perfect refuge.

The fish is obviously immune to the potentially lethal stings of the anemone, and science has yet to agree how this can be so. Simply put, the anemone does not seem to recognize that the fish is an intruder, while other species of fishes and other animals are targeted and attacked instantly upon contact.

One theory is that the anemonefishes have an inert mucus coat that makes them undetectable to the host anemones. A competing theory is that the fish gradually acclimates itself to the anemone, approaching cautiously, brushing the tentacles lightly and nibbling at them before diving in. Somehow it adjusts its mucus coat, either by adapting chemical signals from the anemone itself or by changing its own chemical signature. (Tiny anemonefishes that settle with an anemone shortly after their metamorphosis from larva to fish seem to have almost automatic protection and seem to need no acclimation.)

Whatever the mechanism, it works, and the fish repays its host by bringing nutrients and by spunkily driving away certain butterflyfishes that are known to prey on anemones. ■

Social setting: a Clark's Anemonefish (*Amphiprion clarkii*) with its Corkscrew Tentacle Sea Anemone (*Macrodactyla doreensis*). The anemone may host a pair of anemonefishes or a trio (a breeding matriarch, her male mate, and a sexually immature subadult). Very large anemones may shelter even larger social groupings, especially when a group of juveniles happen to settle into the same anemone.

Damsels in hiding: Juvenile Threespot Damselfish (*Dascyllus trimaculatus*) find temporary safety with a Magnificent Sea Anemone (*Heteractis magnifica*). Closely related to the anemonefishes, damselfishes may also form associations with anemones while the damsels are relatively young and defenseless. These juveniles will mature into pugnacious adults that no longer need the protective cover of the anemone.

Crustaceans in residence: Spotted Porcelain Crabs (*Neopetrolisthes maculata*) live freely among the hair-trigger stinging tentacles of a carpet anemone (*Stichodactyla* sp.). These attractive little crabs are filter-feeders with long antennae and are closely related to the squat lobsters. They may confer some protection to the anemone, which accepts their presence and even allows them to enter its mouth without a typical retraction or capture reaction.

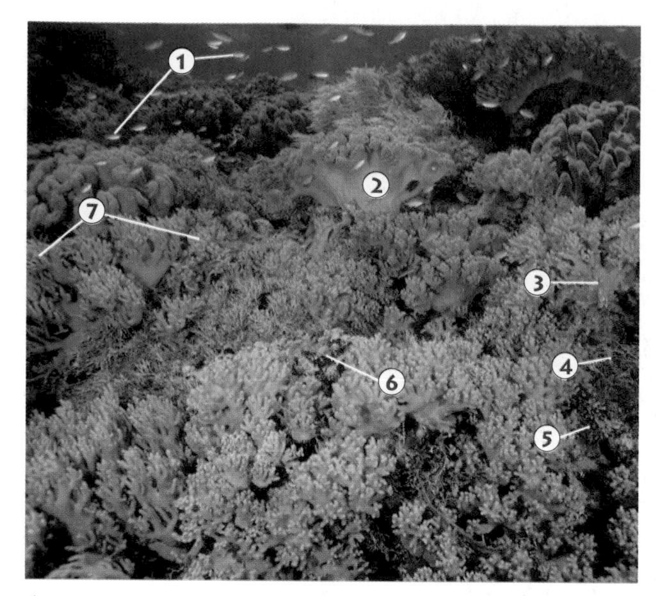

1 Damselfishes (*Chromis* sp.)
2 Leather coral (*Sarcophyton* sp.)
3 Hydroids (*Lytocarpus* sp.)
4 Calcareous algae (*Halimeda* sp.)
5 Cup corals (Dendrophylliidae)
6 Encrusting coralline red algae
7 Soft coral (*Sinularia* sp.)

UNSEEN SKIRMISHES

Corals may cover up to 85 percent of the hard substrate of a reef, and the ability to occupy a piece of this turf and defend it is highly evolved in these animals. Many corals will fiercely nettle any encroaching invertebrates, extending stinging sweeper tentacles to keep competitors at a safe distance. Other corals simply outgrow their foes, eventually overshadowing them and cutting off their life-giving sunlight.

The soft corals shown opposite have been found to possess highly toxic compounds that are believed to be useful both in fending off predators and in keeping encroaching corals from growing too close. In plants, this is known as allelopathy, or the ability to repress or destroy nearby plant competitors. Some of these biochemical weapons are of potential interest to medical researchers. Living together, at least among reef creatures, is not always what it appears.

Field of soft corals: this lush aquascape suggests the abilities of various species to coexist on the same section of reef. Beneath the beauty, however, lies a competitive battleground that science is only starting to understand.

Scavenger: bristleworms are active foragers for dead or dying material, which attracts them by emitting chemical cues. This fireworm (*Chloeia* sp.) is shown in the act of ingesting a dead fish. Some bristleworms scavenge by day, others nocturnally, and some species also attack live corals, clams, and other invertebrates. They are found in great numbers in reef rubble and under loose rock.

Sticky fingers: a common sight in the wild and in reef aquariums, the long feeding tentacles—up to 1 m (3.3 ft.)—of a terebellid or spaghetti worm (Terebellidae) reach out in search of bits of detritus. The threadlike appendages, covered with fine, hairlike cilia, attach to food particles passing by in the water column and funnel them back to the mouth of the buried worm.

Sand processor: many large sea cucumbers, such as this Pineapple Sea Cucumber (*Thelenota ananas*) scour soft substrates, ingesting detritus, small invertebrates, and sand—all of which is passed through the animal's gut to extract all possible nutritive value. The large, spiny protuberances are actually soft to the touch.

Castings: evidence of the prodigious appetite of an acorn worm (Hemichordata), this mound of ropelike fecal matter is actually mostly sand, extruded in a thin casing of mucus. The species responsible may be the Giant Acorn Worm (*Balanoglossus gigas*). Sea cucumbers also produce similarly impressive fecal casts.

Grazing star: a Horned Sea Star (*Protoreaster nodosus*) glides over a sandy bottom, scavenging algae, dead plant and animal material, and possibly sponges. Other species of carnivorous sea stars seek buried mollusks and other prey.

KILLER ANATOMIES

Sharks tend to come to mind first when the subject of deadly marine animals is raised, but many invertebrates and fishes have evolved feeding methods that are every bit as chilling and lethal—albeit on a much smaller scale.

Cnidarians include the hydroids, jellyfishes, sea anemones, and corals, all possessing **cnidocytes**, specialized cells that contain stinging structures commonly referred to as nematocysts, to capture their prey. Nematocysts are like tiny poison-tipped spear guns that can be fired at will when a food source is detected. Hydroids can discharge their nematocysts in 3 milliseconds when they detect a food source. Upon discharge, the stinging cells penetrate the prey, often injecting a protein toxin that paralyzes the unfortunate victim until the tentacles of the predator can grasp it and transport it to the mouth.

Echinoderms include the sea stars and sea urchins, both of which have extreme feeding methods. The notorious crown-of-thorns sea stars evert their stomachs to engulf their prey and dissolve it with gastric enzymes. Other sea stars feed on bivalves and use the tube feet along their arms to force the shells apart. Only a crack is necessary, as the asteroid can evert part of its filmy stomach through an opening as narrow as one-tenth of a millimeter. Within minutes, the stomach enzymes have started to reduce the clam or oyster to a soupy, absorbable texture in its own shell.

Polychaete worms include some notorious raptorial feeders that lunge out and grab prey with complex, heavily armed jaws that have served as models for certain horror film creatures. One particularly large species of eunicid worm is nicknamed the Bobbit Worm and feeds nocturnally. It can exceed 3 m (9.8 ft.) in length and has five sets of jaws that are capable of grabbing and ingesting a fish. Given its size and habit of shooting out of the sand to attack, this rather fearsome worm is enough to give pause even to normally fearless divers (see photograph, page 68).

Marauding herbivores: schooling parrotfish (*Scarus* sp.) overrun a patch of reef, an intimidating swarm that subdues territorial damselfishes that might prove pesky to a single intruder. Many parrotfishes use their powerful beaklike fused teeth to rasp algal film from rock, often taking a portion of rock with each bite, while others eat macroalgae and some ingest live coral. Each large parrotfish may grind out as much as a ton of coral sand each year.

TOOTH & CLAW

Just as hunters and harvesters need tools to accomplish their tasks, marine fishes and invertebrates rely on specially adapted teeth and claws when they feed.

Parrotfishes are mild-manner herbivores, but they have a set of formidable teeth that are fused into a beaklike oral tool strong and sharp enough to bite and crunch the very hard skeletons of stony corals.

Box crabs (Family Calappidae) have massive claws that have no trouble dispensing with mollusk shells to get at the soft body parts. Other crabs and shrimps have claws of unequal size that are adapted for specialized feeding techniques. The smaller claw is used for cutting, while the larger one is more adept at crushing prey. Some brachyuran crabs are known to seize a mollusk, search out the weakest point of its shell, and then apply adequate claw pressure to that spot. Stone crabs, for example, can exert a claw pressure of up to 14,000 pounds per square inch.

Alpheid shrimps, often called pistol or snapping shrimps, also have unequal claws. The larger claw is used for cracking open small bivalves, stunning fishes, in territorial displays, and as a defensive mechanism. The loud cracking sound the shrimp makes is actually the explosive pop of a bubble of air in the claw joint. A flash of light has been reported to accompany this micro-blast of energy release.

Mollusks, especially the gastropods, have a remarkable feeding structure called the **radula** that resembles a series of sharp, toothlike scrapers arranged on a conveyer belt. The radula allows many mollusks to graze efficiently, rasping algae from hard substrates. Carnivorous species use it to grasp, bite, rip flesh, and bore through shells. Carnivorous opisthobranchs, for example, use their radula like a rasping tongue to scrape away the living tissue of their soft-bodied prey. ■

Rasping specialist: An Epaulette Surgeonfish (*Acanthurus nigricauda*) has modified mouth parts that allow it to scrape highly nutritious algal film and diatoms from rocks. Brushlike teeth are common among members of this family (Acanthuridae), but grazing habits differ among the species, some browsing on leafy macroalgae and others seeking out detritus and diatoms on sandy bottoms.

Lawnmower Blenny: an industrious grazer on filamentous algae, the Jeweled Rockskipper (*Salarias fasciatus*) has comblike teeth for currying green growth off hard substrates. Reef aquarists who keep this fish to help control nuisance algae growth have dubbed the species the "Lawnmower Blenny." Some other combtooth blennies graze on coral polyps.

Stout beak: the fused teeth of this female Redlipped Parrotfish (*Scarus rubroviolaceous*) form a beak that both gives the fish its common name and allows it to crunch up dead coral to get at the algae growing on it. Some species grind up live coral to extract the algal cells in the coral polyps.

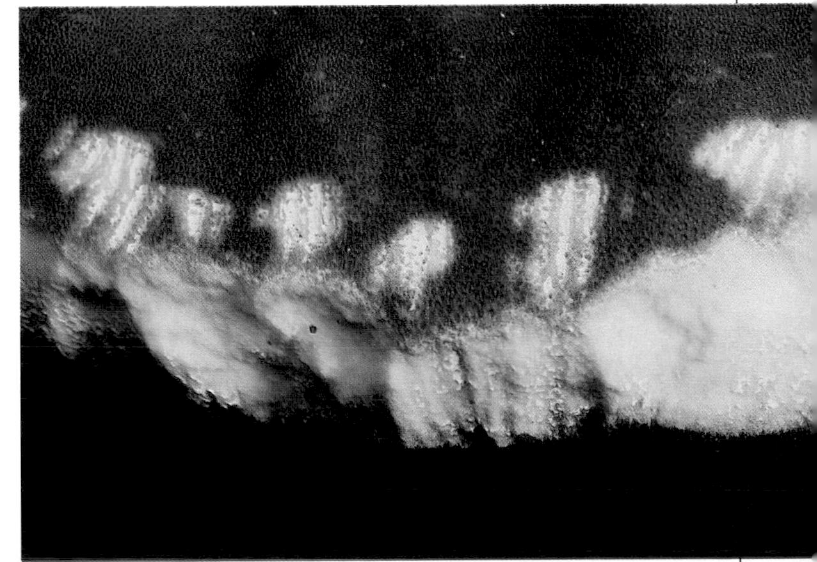

Parrotfish scars: telltale beak marks left on coral by a feeding parrotfish. These fishes take in substantial amounts of coral—both alive and dead—and sediment each day, grinding it up in a specialized pharyngeal mill at the back of their throats. The pulverized material is defecated after passing through the fish's digestive tract.

HUNTING TOOLS

Among the predatory fishes are many that have evolved exquisitely sophisticated anatomies to find and capture their prey most efficiently.

Many of these predators have large mouths with strong teeth and muscular jaws for grabbing and holding prey. Additional teeth, called pharyngeal teeth, are located along the gill arch near the back of the mouth, or throat, of the fish. The pharyngeal teeth chew the prey while the front teeth grasp it and keep it from escaping. Pharyngeal teeth also help prevent the prey from inadvertently escaping through the gill slits. Some predators have extra teeth on the tongue or along the roof of the mouth to aid in swallowing. Others have teeth that allow the prey to move only in one direction—toward the esophagus.

Additional modifications that aid carnivorous fishes include copious mucus production to facilitate swallowing large, irregular, or bony prey. Some fishes swallow their prey whole with the help of an extensible esophagus that moves large prey to the stomach where strong acid secretions digest it. Fishes that swallow their prey whole often have expandable stomachs that can, in some cases, accommodate prey even larger than the predator itself.

Soldierfishes, squirrelfishes, cardinalfishes, and other nocturnal strainers are often red, making them more difficult to see at night, and have large mouths and eyes for improved night vision. Studies have shown that even with their improved vision, night feeders tend to target and eat larger plankton than the daytime planktivores, perhaps due to the visual difficulty of locating prey at night. ■

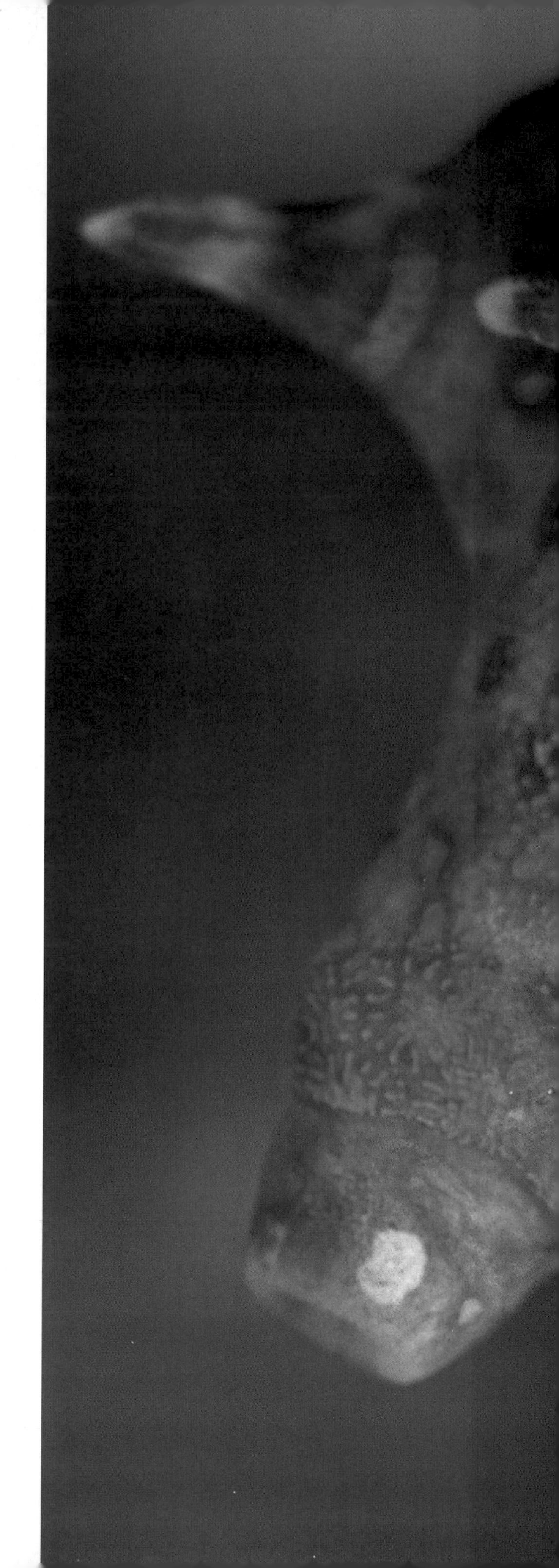

Puff artist: the Thornback Cowfish (*Lactoria fornasini*) displays characteristic horns and a specialized mouth that it uses to jet puffs of water into sandy substrates to expose buried prey—primarily benthic invertebrates.

Raptorial predator: a mantis shrimp (Gonodactylidae) (**1**) feeds on its victim, having darted from its burrow to dispatch a Banded Coral Shrimp (*Stenopus hispidus*) heavy with green eggs (**2**). Normally folded close to the shrimp's body, these lethal "arms" can be opened with tremendous force and speed—in less than 4 milliseconds.

Opportunistic feeders: normally herbivores that graze on green algae and seagrasses, these Mottled Spinefoot Rabbitfish (*Siganus fuscescens*) gang up on an unidentified jellyfish (Scyphozoa). Many plant-eating fishes will also feed on meaty targets whenever opportunities present themselves. Most fishes avoid the stinging jellyfishes.

Crusher: adult Yellowtail Coris Wrasse (*Coris gaimard*) forages over the reef, seeking out crustaceans, mollusks, and even sea urchins, crushing hard shells whenever necessary with its strong teeth and muscular jaws. The huge family of wrasses (Labridae) includes a range of carnivores large and small, as well as many planktivores that pluck their prey from the water column.

Obligate predator: the stunningly pigmented Harlequin Shrimp (*Hymenocera picta*) feeds exclusively on sea stars and is considered to need this specific prey to survive. The shrimps have been observed eating sea stars arm by arm, avoiding the central disc that contains the star's vital organs, and thus keeping the victim alive to provide another day's fare.

Shadowing behavior: Bluefin Jacks (*Caranx melampygus*) "shadow" a large Napoleon Wrasse (*Cheilinus undulatus*), using the large fish as a screen that allows them to sneak up on prey. Trumpetfishes often use this same hunting tactic.

PREDATORY TACTICS

Carnivorous reef fishes can be classified into several general groups based on their distinctly different methods of predation.

Hunters stalk their prey—sometimes exercising what seems to be tremendous patience—before grabbing it. Most are able to summon bursts of speed or lightning-fast reflexes to finish the hunt. Examples include barracudas and trumpetfishes.

Chasers simply outswim their prey with speed and endurance. Examples include reef sharks and jacks.

Ambush predators lie in wait, settling somewhere and waiting for their desired prey to pass by before grabbing it with explosive energy. Examples are groupers, eels, and hawkfishes.

Anglers are fishes that actually lure their prey, by baiting them with a decoy—a specialized anatomical appendage that resembles the prey's food. When the prey moves in to swallow the bait, the predator swallows the prey. Examples include the frogfishes and stargazers, the later having a lure inside the lower lip.

Benthic foragers swim slowly over the reef substrate, picking off prey items, such as crustaceans, that are spotted on the bottom or in hiding places. Examples include triggerfishes, eels, and snappers.

Grazers also make their way over the reefscape, seeking out sessile invertebrates such as sponges, tunicates, and corals. Examples include angelfishes, butterflyfishes, and Moorish Idols.

Grovelers locate live prey by rooting in soft substrates to find worms, crustaceans, and other buried prey. Examples include goatfishes and some gobies.

Pickers are the planktivores that hover in the water column and selectively target certain prey items that meet their criteria for size and shape. Examples include the anthias, damselfishes, and cardinalfishes.

Cleaners establish themselves at permanent "stations" where other fishes come to be groomed of parasites and dead tissue. Examples include certain small wrasses, as well as juvenile angelfishes, and others. ■

NUTRIENT CYCLING

One of the great mysteries of coral reefs is the question of how these dense concentrations of biodiversity can exist in seemingly sterile waters.

Traditional terrestrial biologists usually find the highest concentrations of plant and animal life where there are obvious sources of nutrients: fertile valleys, rainforests, and river deltas.

A coral reef exists in clean, clear waters with virtually undetectable concentrations of dissolved nutrients. As explained on page 44, the symbiotic relationship of stony corals and their zooxanthellae uses the blazing tropical sun and dissolved minerals to build reef structures at a tremendous pace.

The explanation for how the on-going energy balance on a coral reef is maintained in perpetuity is much more complex. Living reef organisms generate organic compounds in the form of body wastes and from the decomposition of their bodies. Marine aquarists understand this implicitly: in a small closed system, nitrogen, phosphate, and carbon compounds can build up at a steady pace. Captive reefs demand special filtration and husbandry techniques to remove such wastes.

On natural coral reefs, the inhabitants act as biofilters and rapidly extract the available nutrients. The simplest example is ammonia (NH_3), excreted by animals as a dissolved, invisible compound, and one that is actually quite toxic to aquatic animals if it is allowed to concentrate. In what is known as the **nitrogen cycle**, certain types of **nitrifying bacteria**—colonizing rock and sand substrates—first metabolize the ammonia to nitrite (NO_2). In turn, other bacteria feed on the nitrite, turning it into nitrate (NO_3). Nitrate can be absorbed and metabolized by algae, corals, and mollusks. These, in turn, are eaten by other animals and the cycle begins again.

In a word, the answer to the remarkable energy balance on the reef is simple: recycling. ■

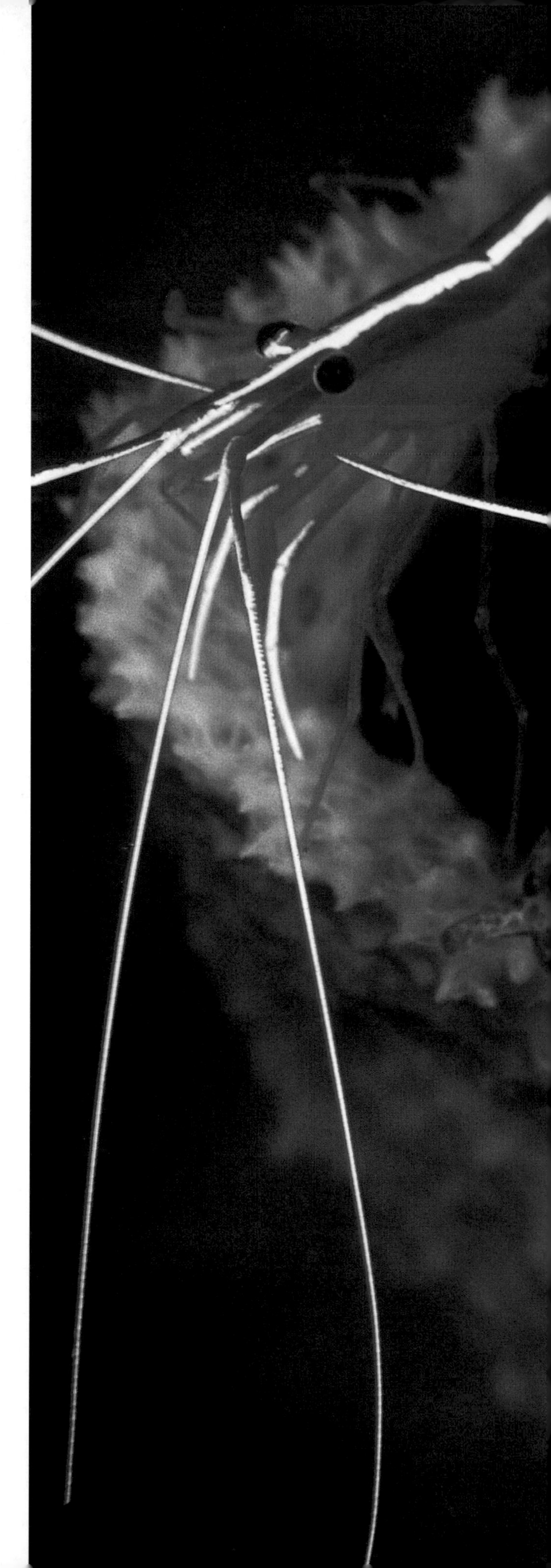

Feeding strategists: a Barredfin Moray (*Gymnothorax zonipectus*) that is a ranging carnivore reposes in a filter-feeding sponge to be groomed of parasites and bits of diseased flesh by Common Cleaner Shrimp (*Lysmata amboinensis*).

transparent and hard to see, even with the magnifying effects of the lens. I couldn't recall having seen photos of juveniles anywhere and the thought of being the first to photograph them really thrilled me. I was more than a hundred feet deep, visibility wasn't great, and a cloudy sky made it seem darker than normal. In my excitement, I forgot about the adult pair and shot the entire roll on the juveniles. That night I developed the film. Imagine my surprise when every frame contained, not a juvenile Ornate Ghost Pipefish, but a needlelike shrimp, Tozeuma armatum, *another symbiont of black coral. Once again, I had a first-hand lesson in just how effective reef animals can be in evading detection and protecting themselves from the attention of predators.*

THE SEA IS NOT FOR THE FAINT OF HEART. It is an unforgiving environment that poses constant challenges to the animals living there. They can never exercise enough caution in their day-to-day lives because danger lurks everywhere. It takes many forms and predation is foremost among them. Although marine animals remain alert and on guard at all times, vigilance does not guarantee their safety.

Predation is a danger all marine animals face, and many anti-predation tactics have evolved. Not being eaten is a matter of survival for all animals, save perhaps the apex predators such as large sharks. As the biggest danger on the reef, predation has given rise to an untold variety of physical and behavioral modifications. Spawning often takes place after dark when the risk of predation is lower. Juveniles and larvae settle out of the plankton at night for the same reason. Many corals and other filter and suspension feeders are active only after dusk when there is less risk.

Juveniles and settling larvae face their biggest challenge in avoiding predators, because their small size and inexperience make them prime targets for aggressive actions. Annual fish mortality estimates are as high as 78% for newly settled larvae and juveniles on the reef, versus 21% for adults, although these rates vary widely depending on time of year, location, predominance of certain species, and many other factors. Mortality rates for larval and juvenile invertebrates are much higher. In some cases, millions of eggs must be produced for only a few to reach maturity. As a counterbalance to the hazards of this dangerous time, young animals have developed their own sets of defensive responses. Some are behavioral, such as settling on the reef at night; others are morphological or related to appearance, such as having distinct juvenile color patterns.

Intruders frequently pose a danger to marine life by seeking to invade the territory of another for food, access to mates, shelter, or as a show of force. Whether it is the turf, the harem, or the nest, protecting the home territory can be a taxing activity.

Some of the more common protective mechanisms found in coral reef animals are:

Hardened skeletons and shells: many marine animals have evolved tough exteriors that discourage or defuse predatory attacks. The calcium carbonate exoskeletons of the crustaceans are effective safety shields against some enemies, while certain fishes, such as the seahorses and boxfishes, have reinforced anatomies to deter predators. Many of the mollusks, of course, have rock-hard shells that provide stout protection from predation and from being buffeted by heavy wave action.

Spines and stingers: stout, needle-sharp weaponry is common among reef fishes and invertebrates. In some cases the spines are simply pointed and forbidding, in others they are loaded with venom. The surgeonfishes, for example, have a scalpel-like "tang" on the caudal peduncle (base of the tail), for use as a weapon. Tiny stinging organs known as nematocysts are potent tools used by both corals and anemones to drive away danger and to kill prey items.

Bristles: some worms have irritating, hairlike bristles that break off in the skin of a predator. Fireworm bristles, or **chaetae**, are sharp, hollow, and filled with venom that is released when they break off. Some urchin spines, or **setae**, also contain venom.

Spicules: these glasslike shards are analogous to spines, but embedded within the living tissue of the animal that needs protection. Sponges and many soft corals have spicules, often razor-sharp, that make them unpalatable to all but the hardiest predators, like turtles and a few butterflyfishes and angelfishes.

Protective coloration: this may involve having a cryptic appearance, mimicking the look of another animal or of the reef substrate, or exhibiting warning coloration to keep predators away. Some protective colors in corals, giant clams, and sea squirts are designed to screen out the lethal effects of intense ultraviolet light.

Nighttime repose: tucked into a mucous sac (**1**) of its own making, a Bicolor Parrotfish (*Cetoscarus bicolor*) sleeps with a sense of security. The temporary cocoon it secretes around itself may disguise the presence of the fish from smell-guided hunters such as large moray eels. The sac may also give the fish an early warning if a predator touches it.

Chemical defenses: plants and animals have an arsenal of defensive toxins and foul-tasting compounds that help them compete and avoid being eaten. Many animals exude mucous coats that make them either unpalatable or difficult to detect (see image above).

Sessile animals, such as the sponges, corals, and sea squirts, are vulnerable to being grazed upon by passing fishes and motile invertebrates. They have internalized their defenses with spicules and chemical repellents that predators have learned to avoid. Many sea cucumbers exude a toxic mucus, others have skin that falls off when disturbed, still others eject **Cuvierian tubules** that entangle predators in a sticky mass of strings.

Flamboyant display: some fishes and cephalopods, such as cuttlefishes and squids, make good use of their colors and fin size when threatened. At times, the best way to run off an intruder is to make him think you are bigger and more threatening than is actually the case. Intensifying colors and expanding fins and appendages are one way of doing this.

Bravado: some fishes seem to use blind fury when threatened, attacking until the intruder backs away. Damselfishes, for example, are sometimes credited with being the meanest fishes on the reef—ounce for ounce. Despite their modest size, they will bravely race out and nip at invading fishes and divers alike. Triggerfishes, which have the teeth to back up their threat displays, are vigorous defenders of their nests and will not hesitate to attack and bite any invader.

Survival is of immediate concern to all marine animals, and each has its own unique tools or behaviors for responding to threatening challenges. Marine biologists are still discovering new and astonishing defenses used by plants and animals of the reef, some of which are expected to have great potential for use in medical and other scientific research. ■

Mass of spines: Crown-of-Thorns Starfish (*Acanthaster planci*), a voracious coral eater, bristles with formidable armor that discourages most attackers. Note stony coral (**1**) it is feeding on.

ARMORED BODIES

Among the most obvious—and often most impressive—defensive adaptations of reef animals are external armaments. The spiny lobsters are classic examples, with a rigid calcium-carbonate **exoskeleton** festooned with sharp spikes. Many echinoderms, such as sea urchins and sea stars, bristle with needlelike spines that can pierce soft tissue and often deliver a dose of venom that may continue to sting for days. Scorpionfishes have venomous spines that they can erect and thrust at approaching predators. Other notable fish groups with venomous spines include the stingrays, waspfishes, and stonefishes, as well as barbel eels, weeverfishes, stargazers, and rabbitfishes. Venomous spines are commonly located in the dorsal, pectoral and/or anal fins, or along the head or shoulder of the fish.

Some fishes have **bony body armor**, notably the sea moths, pipefishes, and seahorses. The platelike covering is made up of scales that have been modified and fused into armor, making them generally unpalatable. The peculiar-looking boxfishes have a protective bony carapace that is unappealing to predators.

Interestingly, many armored animals also carry defensive toxins, making them more intimidating. However, as many species evolved toxicity, a few others evolved a resistance to them. Lizardfishes eat sharp-nosed puffers regularly, groupers eat porcupinefishes, and eels eat various venomous scorpionfishes. These predators may not be immune to the toxins of each other's prey, but over time they have developed resistance to the toxins of their own targets. The ability to overcome anatomical armor and then to resist the toxic nature of certain species defines a niche for the extraordinary predators that can thwart such defenses. ■

Deadly beauty: the Peacock Mantis Shrimp (*Odontodactylus scyallarus*), dubbed the "thumbsplitter" by fishermen, has clublike raptorial appendages (**1**) that can be unfolded with amazing speed to smash their predators. Other mantis shrimps, named for their resemblance to the common garden praying mantis, have fingerlike appendages designed to spear their victims.

Stay away: one of the most vividly pigmented of the reef invertebrates, the Peacock Mantis Shrimp (detail of specimen at left) sports a hardened carapace that suggests a psychedelic coat of armor. Crustacean flesh is a target of many marine predators, and the bright warning colors, calcified exoskeleton with spines, and smashing weaponry of this animal combine to give it significant protection.

Touch me not: fireworms (*Chloeia* sp., pictured), also known as bristleworms, are a diverse group of polychaetes that are protected by a profusion of sharp, hollow setae (**1**) that resemble fine tubes of glass loaded with a stinging toxin. These bristles break off easily and produce a lasting burning sensation in the tissue of would-be attackers. Vivid colors signal a clear warning.

Unpalatable mouthful: bristling, venomous chaetae (**1**) and fleshy, paddlelike parapodia (**2**) cover the dorsal surface of a fireworm (detail of specimen at left). Fishes and other invertebrates feed heavily on marine worms. Species such as this that venture from hiding on their own scavenging missions typically sport both mechanical and chemical defense systems.

MULTIPLE DEFENSES

Curiously, some reef animals have evolved not one, but often two or three lines of defense—often when any single one of them may seem entirely adequate.

The small commensal shrimp pictured here, for example, have a protective exoskeleton, prominent pincers, coloration that effectively camouflages them with their surroundings, and a habit of living out their lives among the spines of a highly protective sea urchin. Crustaceans such as these put on lively displays involving threatening claw waving and scurrying about when defending themselves.

Pufferfishes and porcupinefishes are similarly "armed to the teeth." Many produce highly toxic substances that permeate their organs. They are repulsive to the taste of most predators. Some have sharp, erectable spines. Their bite can be fierce. Finally, these fishes also deter predators by inflating themselves like balloons, so they appear to be larger than they actually are, and hopefully, too large to fit into the mouth of a threatening predator. By packing a triple whammy, these fishes make up for their slow-swimming nature.

In trying to understand the evolution of a seemingly defensive or protective mechanism in a reef animal, it is important not to underestimate the fierceness of the predatory pressure in such a crowded habitat. At the same time, many defensive features typically have other functions in protecting or assisting the animal during its own food-procuring activities or against environmental stresses. ■

Defense in layers: a pair of commensal shrimp (*Allopontonia iaini*) with their own protective exoskeletons (**1**) live safely in the stinging spines (**2**) of a Fire Urchin (*Asthenosoma varium*). Perhaps coincidentally, this shrimp species bears a very close resemblance to a small squat lobster or galatheid crab (*Allogalathea elegans*) that also bears a tough exoskeleton and pincer claws (**3**).

1
3

Biodisguise: a carrier crab (*Ethusa* sp.) (**1**) bears a heavy load of venom-laced protection in the form of a long-spined sea urchin (*Astropyga radiata*) (**2**) that it transports on its rear legs, just above the carapace. In turn, tiny zebra crabs may live among the spines of the urchin.

EPIBIONTS

Space is at a premium on the reef, and many organisms have to take up residence on the surface of another plant or animal. When organisms settle or attach to the exterior of other living organisms, they are known as **epibionts**.

Decorator crabs are an obvious group of animals that actually encourage the growth of such epibionts. Anemone crabs are hermit crabs that decorate their shells exclusively with anemones, which they use to deter predators. When confronted with danger, the crab retracts into its shell and thrusts the stinging anemones at its antagonist. When this soft-bodied crab is ready to move into a bigger shell, it strokes the anemone body with its claw to encourage it to release its hold on the old shell. When the hold is released, the crab transfers its symbiont to the new shell. Other uses of such epibionts abound (see facing page).

Some of the sessile mollusks, such as the Thorny Oyster (*Spondylus varians*), have roughened shells that offer a perfect substrate for the growth of sponges, calcareous algae, and other epibionts. This veneer of live matter tends to make the oyster disappear into its piece of the reef. Some fishes also seem to use the same tactic, allowing algae to coat or encrust their skin, giving them a perfect camouflage for protection and a convincing disguise to fool their prey. These fishes seem to shed mucus as a protective barrier and sometimes rid themselves of all epibionts. Some scorpionfishes, lionfishes, and waspfishes seem to "jump" out of their skin by shaking off an entire epidermal layer, known as the cuticle, in occasional shedding events. ■

The crab within: defended by stinging hydroids that it has attached to its carapace, this spider crab (Majidae) both camouflages its appearance and effectively signals inquisitive predators to keep their distance. In its own hunting efforts, the crab may create feeding opportunities for the hydroids that are its involuntary partners.

Funky chapeaux: a decorator crab (*Cyclocoeloma tuberculata*) cultivates its own bio-shield, having placed stinging corallimorpharians and noxious sponges on its back. The crab's carapace is covered with tubercles—hence the species name—that provide a textured surface and attachment points for the hijacked items it collects.

Jelly with legs: a decorator crab (*Ethusa* sp.) lives under the umbrella of an Upside-down Jellyfish (*Cassiopeia andromeda*) that it carries with two pairs of specially modified legs. The flattened crab thrusts the jellyfish, which has mild stinging abilities, at would-be attackers to chase them away.

Anemone pugilist: the Boxer Crab (*Lybia tessellata*) carries a small pom-pom anemone (*Bunodeopsis* sp.) in each claw and jabs them at anything it perceives to be a threat. The mobility provided to the anemones allows them to capture food both for themselves and, inadvertently, for their host.

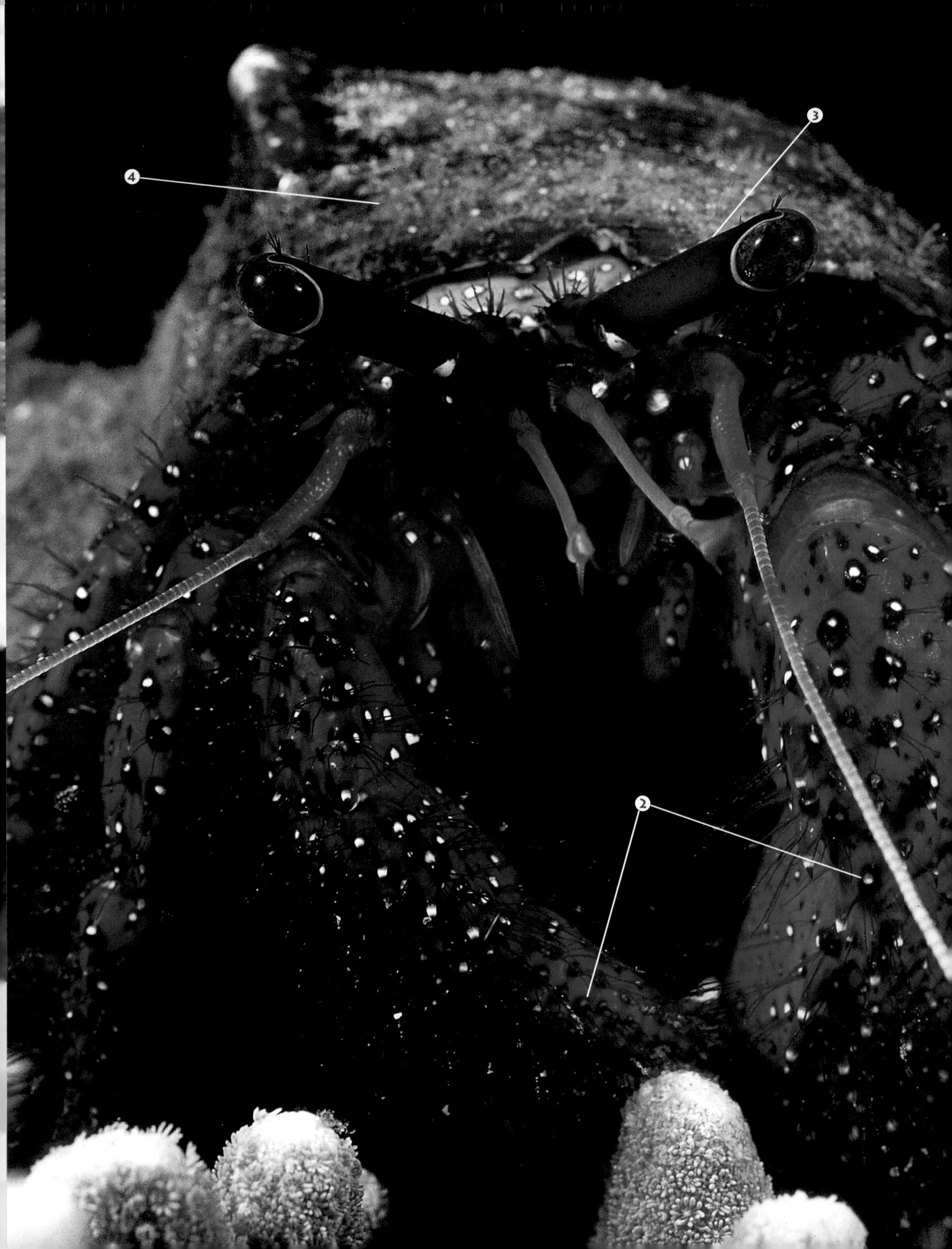
4
3
2

HERMITS & HIDERS

Hiding is a simple but effective anti-predation technique. Many animals take refuge within fissures, nooks, crannies, tunnels, and overhangs on the reef, as well as under coral rubble and sand. By arranging their own periods of activity to coincide with the inactive periods of their predators, many animals lessen their chance of being preyed upon. Eels, for example, have relatively poor eyesight and generally remain within the safety of their lairs during the day. At night, when visibility is reduced and predators are scarce, eels venture out in search of prey that they are able to detect with their other senses.

Some of the parrotfishes hide themselves at night by secreting protective **mucous cocoons**. In order to get to the fish, the predator must first go through the mucus sac, thereby alerting the parrotfish and giving it time to escape. Some observers believe the mucus itself may be distasteful or somehow confusing to hunting species, such as eels, that are guided by smell.

Hermit crabs have evolved a body form in which the abdomen and tail are soft and totally inviting to predation. To protect themselves, these familiar crabs always hide in discarded mollusk shells, for which there is considerable competition—a crab caught out of its shell is soon gone. When one hermit crab seeks to displace another for its larger or superior shell, the aggressor grabs the defending crab with its claw and tries to pull it out of its shell with its greater strength. The defending crab resists this action by retreating as far as possible into its shell to escape the claws of the aggressor. If the aggressor fails, it sometimes picks up the desired shell, crab and all, and drags it around for a while, perhaps hoping the crab under siege will weaken and become easier to displace. Hermit crabs sense when one of them is about to vacate a shell and form a line behind the main aggressor. The order of the line is determined by size and might—the contenders fight for a good position. ■

Portable refuge: the Red Hermit Crab (*Dardanus megistos*) has a prickly exterior (**1**), strong pincers (**2**), and long eye stalks (**3**) to peek out of its shelter in an abandoned snail shell (**4**). Hermit crabs must be able to withdraw completely into their shells for effective protection, and they must find and shift themselves into ever-larger shells as they grow.

Aposematic coloration: this polyclad flatworm (*Pseudoceros ferrugineus*) glides over the reef in search of tunicates that form its highly selective diet. The bright pigments of soft-bodied invertebrates typically signal their unpalatability.

DANGER SIGNALS

Some marine animals are naturally protected from being eaten because they are toxic or distasteful to some or all predators. After the first unpleasant encounter with a noxious animal, the predator learns its lesson and subsequently avoids that animal. To etch the memory of the distasteful experience into the brains of their potential enemies, many marine animals advertise their unpalatability by displaying bright colors or bold patterns as a warning to would-be predators.

This is known as **aposematic coloration,** and animals that display this defense gain protection from most predators. Many carnivorous and omnivorous fishes have come to equate bright, bold colors with distasteful or toxic experiences.

Among the most noteworthy examples of animals that boldly advertise their distasteful or poisonous nature are nudibranchs, polyclad flatworms, and certain fishes. Soapfishes and boxfishes have **skin toxins** that deter predators. They can secrete substances strong enough to kill other fishes kept in the same aquarium. Some soles produce an exudate that repels even sharks, while dragonets secrete a foul-smelling and distasteful, but not toxic, mucus. Some eels and gobies also produce skin toxins. Many of these species have bold, instantly recognizable color patterns that signal foes to steer clear.

Many of the highly venomous sea snakes or sea kraits of the Indo-Pacific have unmistakable alternating bands of black and white. The venom is a **neurotoxin** not unlike that of the cobra. They glide freely over the reef during daylight hours and are unmolested by fishes that somehow know to avoid them. Even humans have an instinctive recoil reflex when exposed to these marine snakes. Occasionally, fishermen fall victim to their bites, usually inflicted when the snakes are caught in their nets. Fatal attacks on divers and snorkelers are virtually unknown. Still, aposematic colors are clear signals that all who interact with Nature ignore at their own risk.

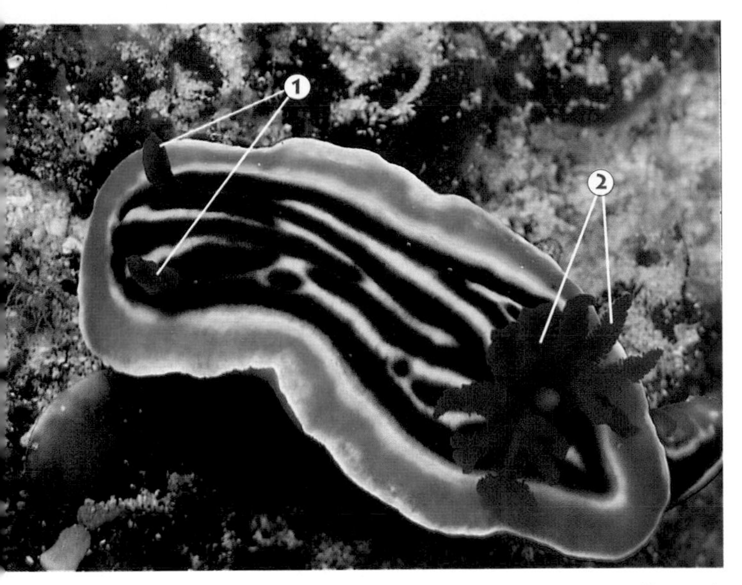

Avoid-me color scheme: this well-known nudibranch (*Chromodoris magnifica*) is a gastropod mollusk that has a shell only during its larval life. It depends on classic aposematic coloration to warn predators that its flesh—or perhaps a mucus shield it exudes—is toxic or extremely distasteful. Note rhinophores (**1**) and gills (**2**).

Noxious beauty: this phyllidiid nudibranch (*Phyllidia* sp.) is a sponge-eating shell-less sea snail that is common throughout the Indo-Pacific region. It is part of a group of nudibranchs with a repellent chemical defense that fishes and hunting crustaceans associate with these blazing colors and learn to avoid.

Bitter memories: one taste of this lovely nudibranch (*Hypselodoris bullocki*) is usually sufficient to teach a predator that it is a bitter mouthful. Scientists believe that the natural toxins produced by reef invertebrates such as this may have future commercial or medicinal applications.

Nematocyst poacher: armed with stinging nematocysts in its fingerlike cerata (**1**), this aeolid nudibranch (*Flabellina rubrolineata*) acquires its weaponry by ingesting hydroids and storing their specialized projectile-emitting cells for future discharge—usually into the soft mouthparts of a fish.

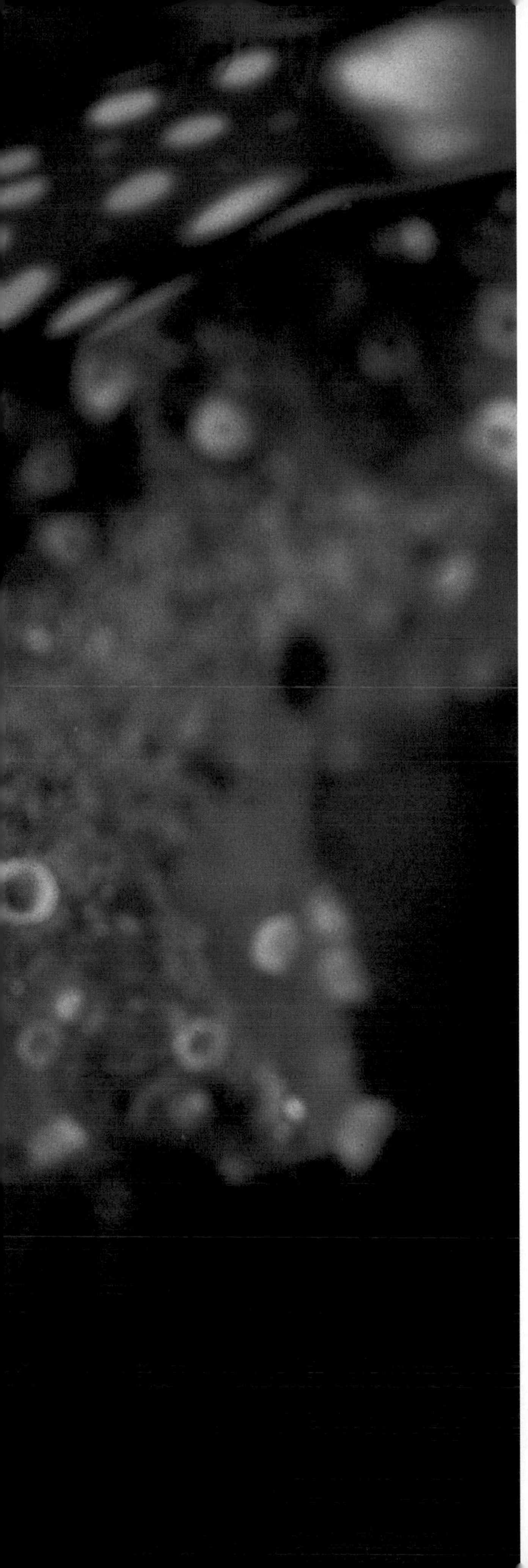

ADVANCED TECHNIQUES

Some protective strategies used by reef animals verge on the extreme and bizarre.

Aeolid nudibranchs, for example, feed on sea anemones and hydroids, somehow having immunity to the cnidarian nematocysts that so effectively sting most other creatures. Nematocysts have a hair-trigger response to touch and a potent sting, but somehow the aeolid nudibranchs are able to eat nematocyst-studded tentacles without apparent harm.

These nudibranchs also transfer and store juvenile nematocysts that mature in their own **cerata**, the characteristic fleshy projections arranged along their backs (see page 153, bottom right). When disturbed, these nudibranchs have their own specialized **cnidosac** cells that can discharge the adopted cnidarian nematocysts to deter predators. In some species, the cerata can also be cast off, wiggling, to distract the predator while the nudibranch makes a hasty retreat to safety. As yet another line of defense, some species resemble the coral polyps on which they feed.

Crustaceans and echinoderms, especially sea, brittle, and feather stars, have the ability to regenerate lost body parts, a definite advantage when dealing with predators and rivals. Many will cast off an arm and sacrifice it to a predator, a practice known as **autotomy**, while the rest of the animal escapes.

Some species of feather stars have "decoy" gonads along their arms. These lures attract and satisfy fishes that normally prey on feather stars, while keeping them away from the real gonads. The decoys are effective, the predators are happy, and feather star fertility is not affected. The sacrificial or cast-off parts, like the self-eviscerated internal organs that sea cucumbers sometimes spew out to escape being eaten, will regenerate in a period of weeks to months. ■

Versatile cloak: a vibrant polyclad flatworm (*Pseudoceros lindae*) is—at first glance—similar to the nudibranchs, but lacks the paired tentacles, or rhinophores, on the top of its head. It bears a distinctive pattern of hues that serve to warn away enemies and perhaps to attract its own kind during mating sessions. Here, it glides over a sponge (*Desmacella* sp.) that hosts numerous tiny jellyfish (*Nausithoe punctata*), seen as tubular white polyps.

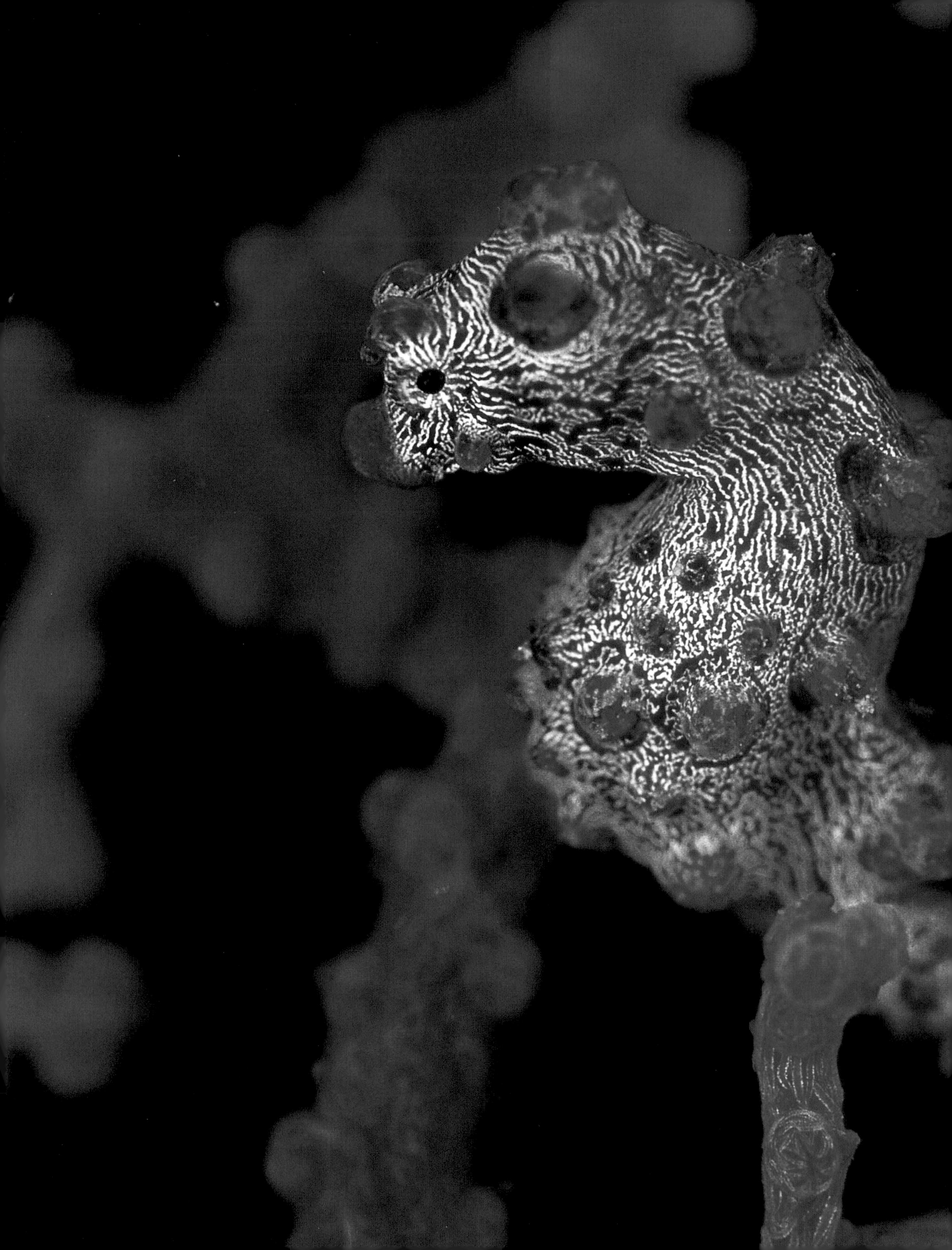

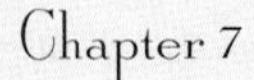

Chapter 7

MASQUERADE

Mimicry and Camouflage

"There is more here than meets the eye."

LADY MURASAKI,
THE TALE OF THE GENJI

For over a decade, my primary occupation was collecting tropical marine sponge samples for medicinal research. The idea was to sample as many different sponge species as possible to provide a broad coverage in any given area. Some species of sponges are known to produce different chemicals in different geographic areas, due, perhaps, to environmental differences, and for me, this meant sampling many of the same species in every area in which I worked. After collecting thousands of samples from a variety of countries, I was pretty attuned to what was a sponge and what was not. I had a reasonably good memory for what I had previously seen and/or collected, and I always got a kick out of finding something unfamiliar to me.

One day, while working off the coast of Borneo, I spotted an unusual pale green sponge. I was thrilled because I had been working in the area for some time and chances of finding a sponge that had thus far eluded me were slim at best. But here it was, a magnificent specimen, unlike any I had ever seen. I quickly

Excellent mimicry: Pygmy Seahorse (*Hippocampus bargibanti*) exhibits remarkable camouflage, living in the branches of a sea fan (*Muricella* sp.) and mimicking even the bumpy retracted polyps of its host. This species' size—often less than 2 cm (0.8 in.)—and astonishing disguise make it notoriously hard to find.

scanned the area for others like it, but there were none obvious. I certainly wasn't going to take the only one, so I would have to search the reef diligently to locate more of this new sponge. I swam over for a closer look and began to examine the surface. It was then that I noticed the eye. It was small and black, but it was definitely an eye. At first I thought that something was peering out at me from inside the sponge. What secrets did this curious specimen hold? I was deep in thought as I continued my examination. When I saw the huge mouth, I flung myself backward and almost swallowed my regulator. My marvelous sponge had turned into a giant frogfish right before my eyes.

THE MARINE WORLD IS FULL OF ANIMALS masquerading as something they are not. In some cases natural selection over millions of years has created animals with appearances that uncannily match other organisms or objects in their environment.

Other animals have the ability to transform their appearances, sometimes in a mere matter of seconds, sometimes over a period of time. These creative illusions can be used for protection, for concealment from antagonists, or for fooling prey organisms that are being hunted. In many cases, the masquerade fulfills all these functions.

Some types of illusory behavior are so specialized that they have been given identifying names that are now synonymous with that type of behavior. The major categories covered here are camouflage, coloration, and various types of mimicry.

Camouflage is the act of disguising or concealing the presence of an individual, either by passive or active means. Cryptic modifications or behaviors give an animal the appearance of being something it is not, like an integral part of its physical environment, in order to escape detection.

Coloration is an important survival and hunting tool on the reef. The uses of colors, patterns, and changes in hue or intensity can be part of camouflage, mimicry, or communication strategies—fear, threats, and sexual overtures. Among the many color adaptations and tactics commonly seen on the coral reef are: **disruptive coloration** (patterns that break up the silhouette), **ocelli** (false eyespots that confuse predators), **flash coloration** (rapid changes in appearance), warning or **aposematic coloration** (bright danger signs), and **diurnal/nocturnal coloration** (day/night shifts in appearance).

Countershading is one of the easiest color camouflages to recognize: fishes that appear light against a light background and dark against a dark background. The dorsal (upper) half of the body is often darker than the ventral (lower) half or underside. The dark dorsal color allows the fish to blend in with either the bottom terrain or the dark blue of the ocean when viewed from above. When viewed from below, the lighter ventral side makes it difficult to see the animal against the surface when daylight filters down through the water. Cartilaginous fishes such as sharks and rays exhibit the most striking examples of countershading.

Mimicry is the advantageous resemblance of one organism (the mimic) to a different organism (the model) in appearance, behavior, or both. Some animals are born physical mimics and closely resemble something else throughout their lives. Others actively behave or shift their appearance in certain ways to create successful deceptions.

Batesian mimicry is a form of biological resemblance in which a noxious or dangerous animal serves as a model for a harmless mimic animal. Many "model" marine animals are naturally protected from predators because they are toxic or distasteful to the predator. After the first unpleasant encounter with a noxious animal, the predator learns its lesson and subsequently avoids that animal—and all others that look like it.

Animals often advertise their unpalatability by displaying bright colors or bold patterns as a warning to would-be predators. This is aposematic coloration and the animals that display it gain protection from most predators. When an unprotected species displays a color or pattern similar to a recognized aposematic coloration, it too gains protection from would-be predators. This is classic Batesian mimicry and it is an effective way for some species to escape predation.

One well-known case of Batesian mimicry involves a toxic pufferfish and a harmless filefish (page 171). The Saddled Filefish, *Paraluteres prionurus,* bears a striking resemblance to the toxic Saddled Toby, *Canthigaster valentini.* Only careful scrutiny of the fins differentiates the two. The skin and tissues of the puffer contain the powerful neurotoxin tetrodotoxin, and its

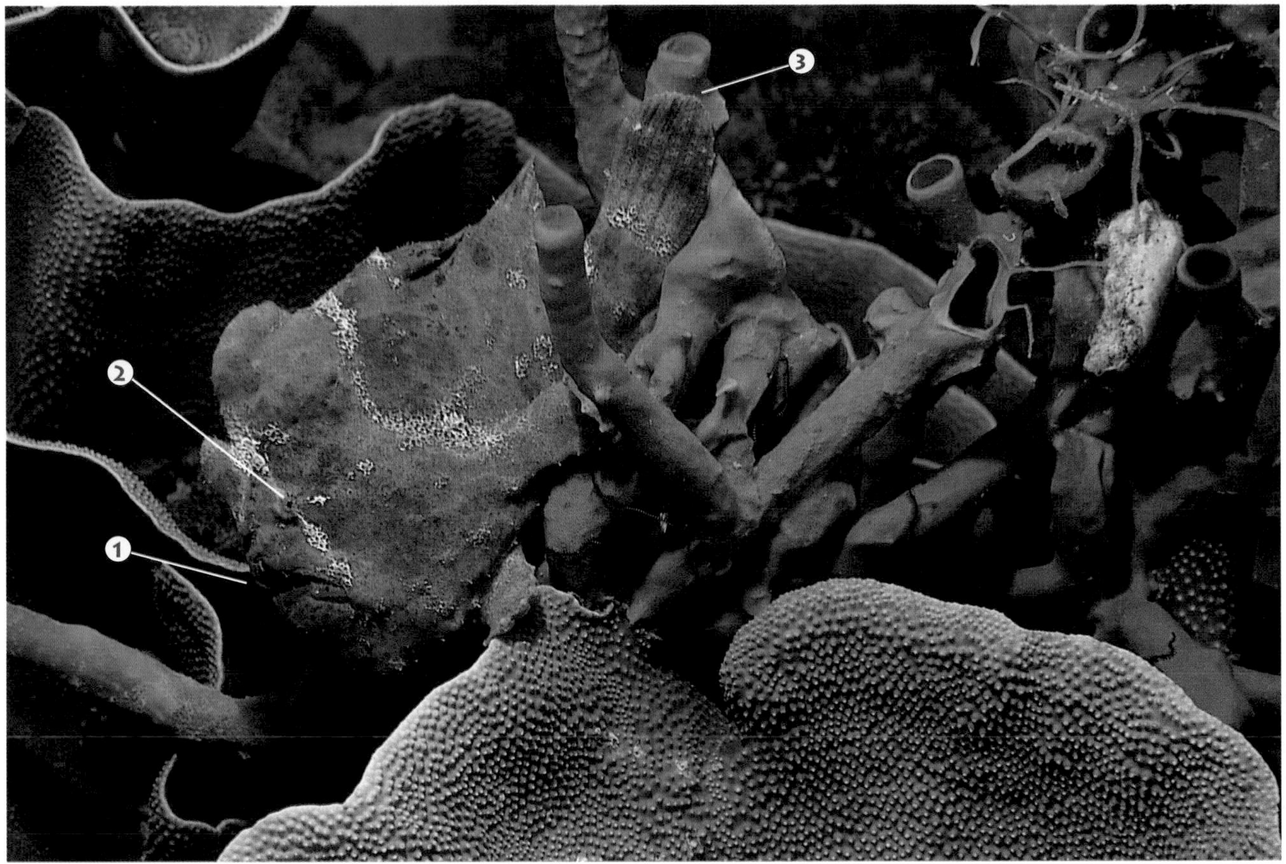

Aggressive or decoy mimicry: a Painted Frogfish or Anglerfish (*Antennarius pictus*) perfectly mimics a green sponge growing amongst a colony of a cup coral *(Turbinaria frondens)* and blue sponges and awaits an unsuspecting fish or crustacean to move within its striking range. Note mouth (**1**), tiny eye (**2**), and tip of tail (**3**).

conspicuous black, white, and yellow coloration warns predators away. The Mimic Filefish has adopted the same coloration so that it, too, enjoys an unusual degree of freedom from predation.

Batesian mimicry works because the number of mimics, or harmless species, is relatively small relative to the number of protected, or noxious, species. If the mimics were to outnumber the truly harmful or unpalatable species, predators would soon learn that the odds of choosing the harmless mimic would outweigh the odds of choosing the undesirable species.

Curiously, quite a few reef fishes are known to employ Batesian mimicry during their juvenile stages only. For example, juvenile Convict Blennies mimic juvenile Striped Eel Catfish in appearance and behavior (see page 173). Both are small, dark brown with white stripes, and move about the reef in tightly packed schools. The Striped Eel Catfish are toxic but the blennies are not. The young blennies thus benefit by getting the same protection as the young catfish because they are so difficult for predators to tell apart.

Müllerian mimicry is the mutual resemblance of two (or more) noxious or dangerous organisms to each other. For example, some polyclad flatworms mimic nudibranchs, and vice versa (see pages 152-153). Both are toxic or at least distasteful. Possibly one is more distasteful than the other, or perhaps both are equally distasteful and their similar appearances and the repellent chemicals they carry reinforce the warning to predators to keep their distance.

Aggressive or **decoy mimicry** as seen in the case of the Painted Frogfish, above, is self-concealment to fool prey organisms or the use of specialized appendages, such as lures (page 117), to bait potential food targets.

This is only a sampling of the wonderful and perplexing world of marine masquerade and illusion. There are many other examples of mimicry and camouflage in the sea. The true test of any deception is whether or not it is detected, and surely there are many successful ruses that have yet to be discovered. ■

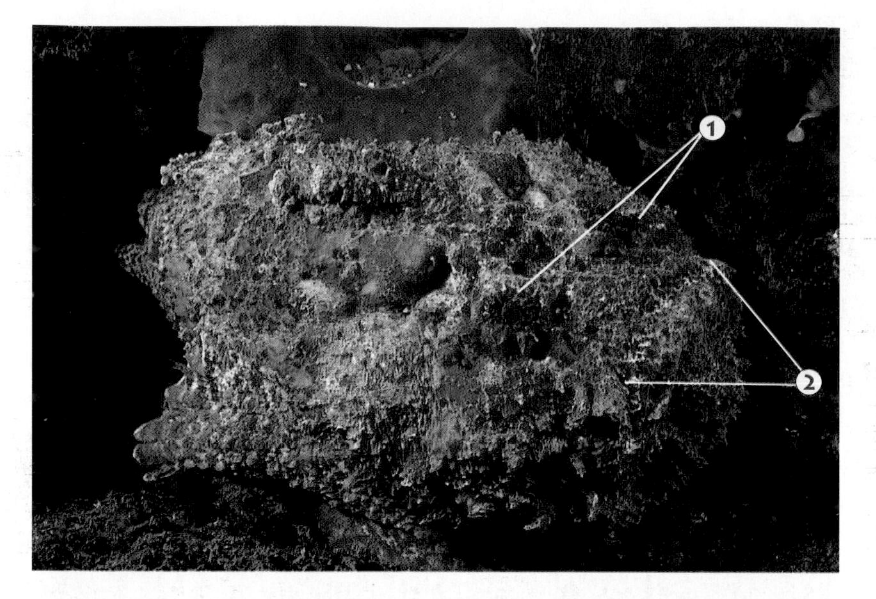

Decoy mimicry: this Reef Stonefish (*Synanceia verrucosa*) appears to be just another piece of reef rock, sitting absolutely motionless and letting its warty skin and the epibiotic growth on its exterior cause it to blend into the surrounding substrate. Surprisingly, the stonefishes also pack a potent venom in the spines of their dorsal and anal fins and have been blamed for serious injury and death to humans who were stung. Note concealed eyes (**1**) and mouth (**2**) of the stonefish.

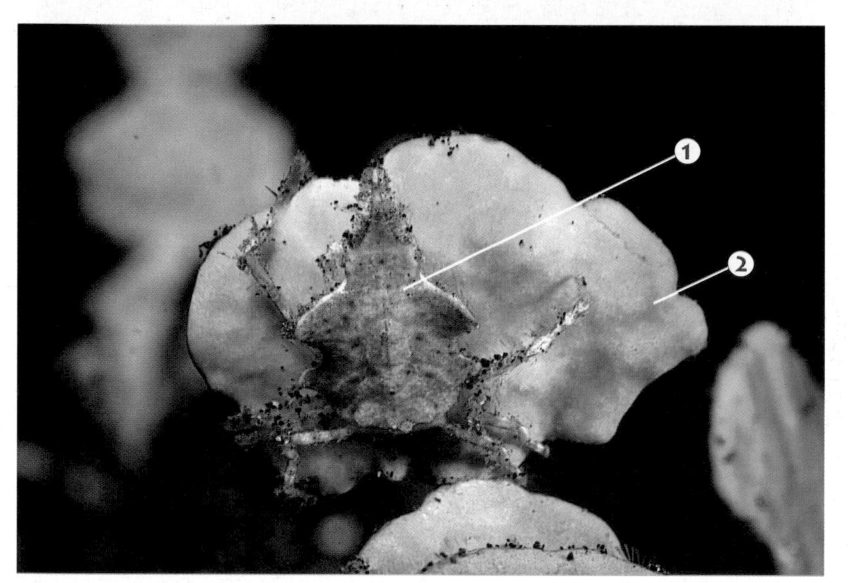

Passive camouflage: Halimeda crab (*Huenia* sp.) (**1**) bears an uncanny resemblance to segments of the green calcareous *Halimeda* macroalga (**2**) that is its frequent host. Examples of such astonishing mimicry are not rare on the coral reef, but finding these animals requires patience and a keen eye.

Active camouflage: this common Short-spined or Collector Sea Urchin (*Tripneustes gratilla*) (**1**) covers itself with pieces of vegetation (**2**), bits of coral rubble, and other found objects to hide itself. Some observers have speculated that the urchin may also use this technique to provide itself with shade when exposed to intense sunlight in very shallow water or to save edible findings for future consumption.

Black coral mimic: the Sawblade Shrimp (*Tozeuma armatum*) (**1**: between pointers) has an extreme elongation of its rostrum (pointed front of its carapace) (**2**), like a branch extension of the black coral (*Antipathes* sp.) on which it dwells.

Concealment: this crinoid shrimp (*Periclimenes* sp.) (**1**: between pointers) is an obligate commensal with feather stars (Crinoidea), taking on the patterns and colorations of its host to escape the attention of carnivorous fishes.

Deadly model: the bold black-and-white pattern of the Banded Sea Krait (*Laticauda colubrina*) is a universal warning on tropical Pacific reefs, associated with an animal whose bite can easily kill a human. This snake hunts with impunity during daylight hours, but is not known as being aggressive toward divers.

Harmless mimic: the Banded Snake Eel (*Myrichthys colubrinus*) is a harmless creature that causes an instinctive avoidance reaction because of its close-enough resemblance to the venomous Banded Sea Krait, at left. A number of other eels, especially the juveniles of several species, are also protected by similar coloration.

Toothy model: the Whitemouth Moray Eel (*Gymnothorax meleagris*) is a known reef predator that is respected by smaller fishes, who learn to associate the general body shape and white-speckles-on-black pattern with danger. Similar variations on this color scheme are used by a variety of different moray eel species. This species is sometimes called the Comet Moray.

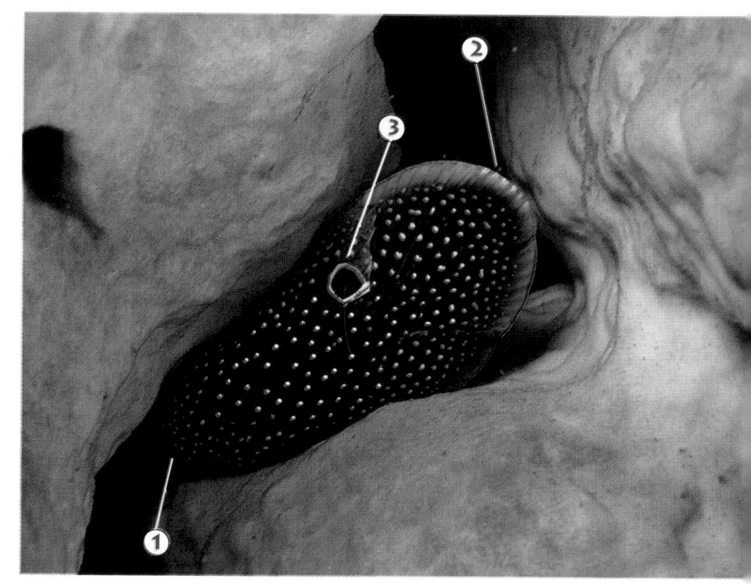

Mild-mannered mimic: the shy Comet (*Calloplesiops altivelis*) casually mimics the Whitemouth Moray Eel with its colors. When threatened, it sometimes assumes a posture suggesting the business end of an eel. The fish angles its head (**1**) downward into a crevice, exposes its tail (**2**) and ocellus, or false eyespot (**3**), and weaves gently like an eel.

Innocent model: a Green Moray Eel (*Gymnothorax funebris*) allows a pair of cleaner wrasses (*Labroides* sp.) to approach and groom it of external parasites and diseased tissue. The eel and cleaners have a ritualized mutualistic relationship, and the wrasses are even allowed to enter the eel's mouth to pick between its teeth.

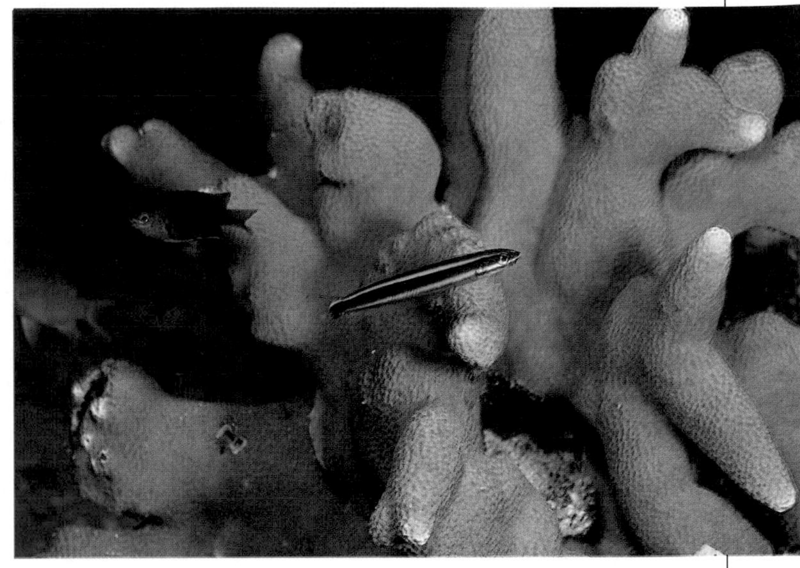

Aggressive mimic: in a case of the wolf hiding in a sheepskin, the Bluestriped Fang Blenny (*Plagiotremus rhinorhynchos*) mimics the cleaner wrasses, at left. Also known as a sabretooth blenny, it is not a beneficial cleaner, but rather uses mimicry to get close to fishes and take small bites of their fins and scales.

Stinging models: a mass of Striped Eel Catfish (*Plotosus lineatus*) juveniles, all bearing venomous spines, swarms over a sandy bottom. Foolish predators or hapless victims that bumble into such a pack may receive hundreds of burning stings.

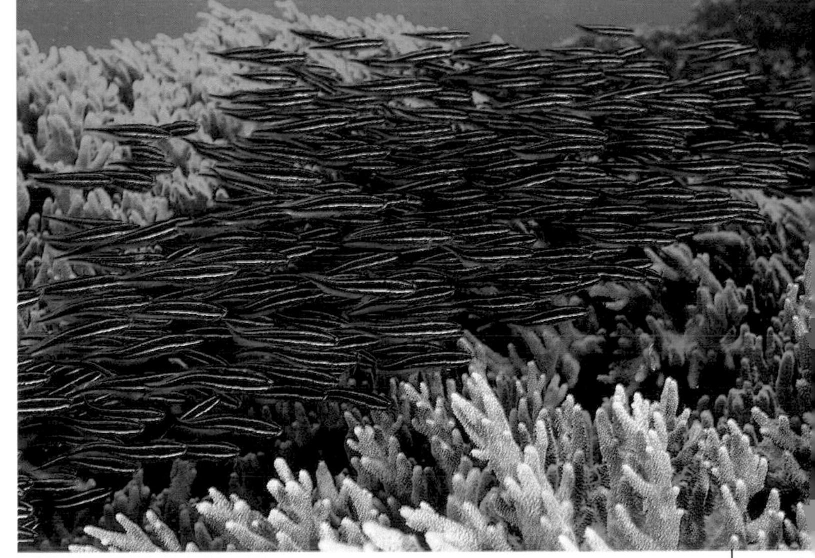

Edible mimics: juvenile Convict Blennies (*Pholidichthys leucotaenia*) that are entirely lacking in defenses form schools that replicate the look of schooling juvenile Striped Eel Catfish, at left. Even a moment's pause by a confused predator may allow such fish to escape.

Master of mimicry: appearing as itself, the so-called mimic octopus is part of an undescribed genus that can use bodily contortions to imitate other reef animals. The octopuses possess highly developed eyes and can distinguish small objects from a distance. Their superior sensory perception and intelligence may work to give them a repertoire of mimic acts.

MIMIC OCTOPUS

The so-called "mimic octopuses" are very special animals that belong in a category by themselves. As of this writing, they are new to science and the subject of growing excitement in the marine biology world. When officially named and described, these animals will belong to a new genus of long-armed octopuses and are expected to have at least two, and possibly four, species.

Their uniqueness lies in a fantastic ability to mimic other marine animals in their environment by manipulating their body, arms, and posture to change their appearance. At least one other species of long-armed octopus has been known to fold its arms back and swim like a flounder. The true mimics, however, have a fascinating repertoire of impersonations that go far beyond flatfish imitations.

True mimic octopuses sit out in the open and rapidly change shape, as if morphing from one animal into another. Their coloration and skin texture do not appear to have the range of manipulation of other octopuses, but their body contortions border on the fantastic: there are reliable reports of these animals transforming into a facsimile of a lionfish, sea snake, jellyfish, flounder, seahorse, mantis shrimp, sand anemone, feather star, brittle star, and stingray, among others (see next two pages of photographs).

When I've observed these animals, they have shown no fear of me. Instead, they go through a series of rapid transformations, morphing from one animal to another, all the while slowly working their way back to their dens. Perhaps they mimicked something in an attempt to scare me away, and when that didn't work, they tried another impersonation, and another, until they either found one that worked or reached the safety of their den. Possibly these octopuses are highly venomous and have no need to fear potential predators. Perhaps a flamboyant warning is enough. Only further study will provide the answers about this unique group of animals. ■

Model anemone: a sand anemone (*Actinostephanus* sp.) with distinctive snakelike arms is covered with stinging nematocysts that most fishes have learned to avoid.

Mimic octopus "anemone": caught in the open, a mimic octopus may go through a series of impressions, including a version of a sea anemone with waving tentacles.

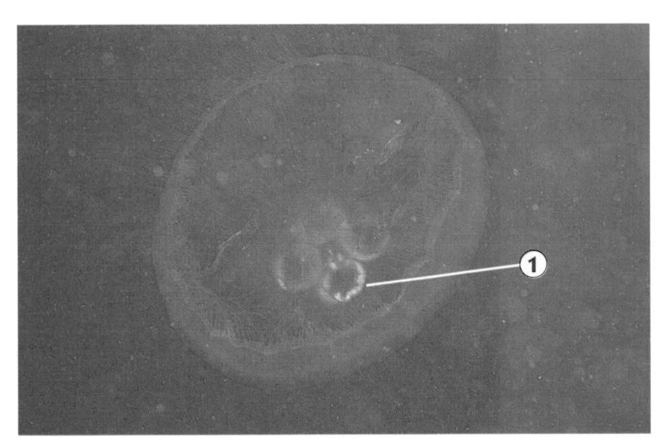

Model jellyfish: a Moon Jelly (*Aurelia aurita*) drifts in the water column. With their stinging tentacles, jellyfishes are avoided by most marine fishes. Note the four rings of gonads (**1**) that are a visual feature of these animals.

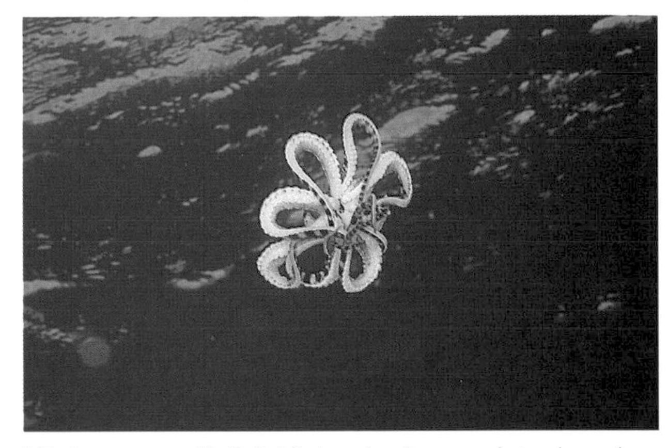

Mimic octopus "jellyfish": bowing its arms into rings gives the floating octopus more than a passing resemblance to a jellyfish with its characteristic white gonadal organs. Such poses are created to confuse threatening predators.

Model flounder: a left-eye flounder (*Bothus* sp.) is a cryptic, bottom-dwelling flatfish. Several different octopus species are known to mimic the flounders.

Mimic octopus "flounder": the remarkable mimicry of the octopus includes imitative body movements that enhance the impression of seeing something other than an octopus.

Model arrow crab: a long-legged arrow crab (*Latreilla valida*) skitters across a bottom of lava sand. Why the octopus seems to imitate such creatures is unknown.

Mimic octopus "arrow crab": here the more colorful "Wonderpuss" species (see facing page) appears to mirror the look of a leggy arrow crab.

Model brittle star: a spiny and not particularly toothsome brittle star (*Ophiothrix* sp.) clambers over a branching staghorn coral (*Acropora* sp.).

Mimic octopus "brittle star": arms constricted, carefully spaced out, and arced give the mimic octopus the look of a brittle star.

Model cuttlefish: a Broadclub Cuttlefish (*Sepia latimanus*) hovers over the bottom. This species is capable of dramatic shifts of pigmentation.

Mimic octopus "cuttlefish": shrunken into the shape of a cuttlefish, this octopus imitates a cephalopod in its boldest color pattern, used to startle predators.

Accomplished hunter: although unidentified, this small, colorful mimic octopus is part of a group of highly unusual mimic octopuses in Indonesia that have begun to attract the attention of marine zoologists. Nicknamed "Wonderpuss," this animal uses its versatile anatomy to capture prey in an intelligent and deliberate manner. In this stance, it is prowling the bottom in search of a prey item, perhaps a fish or crustacean. Note arms in comparison to sequential photos, below.

Erecting the trap: with a target food item under its body, the octopus raises itself slightly and flares the membranes (**1**) between its tentacles, trapping the victim.

Meal in a pouch: sealing all avenues of escape for the animal caught beneath its body and encircled by its arms, the octopus is able to eat without fear of losing its prey.

Müllerian co-mimic: when two unpalatable or feared species resemble each other, it is known as Müllerian mimicry. In this case, a Striped Fang Blenny (*Meiacanthus grammistes*), above, and a Striped Mimic Blenny, (*Petroscirtes breviceps*), below right, are a close match for each other. The fang blenny is mild-mannered but venomous, while the mimic blenny has fangs, but no venom glands associated with them, and is known to bite with very little provocation. Both have nasty aspects, and the similarity of one to the other seems to protect both. Which is the chronological mimic is unknown.

Bates, meet Müller: the Blackstriped Cardinalfish (*Cheilodipterus nigrotaeniatus*) is a harmless Batesian mimic that resembles the two Müllerian mimic blennies shown above and right, that bigger fishes tend to shun.

Co-mimic: the Striped Mimic Blenny, (*Petroscirtes breviceps*) is an aggressive species and a Müllerian mutual mimic with the fish above. It may also serve as one model for the meek little cardinalfish, at left.

Crypsis: Cockatoo Waspfish (*Ablabys taenianotus*) packs a potentially lethal dose of venom in its spines, but it perfectly mimics dead leaves as it lies in ambush. Live algae may cover its skin. Note eye (**1**) and dorsal fin (**2**).

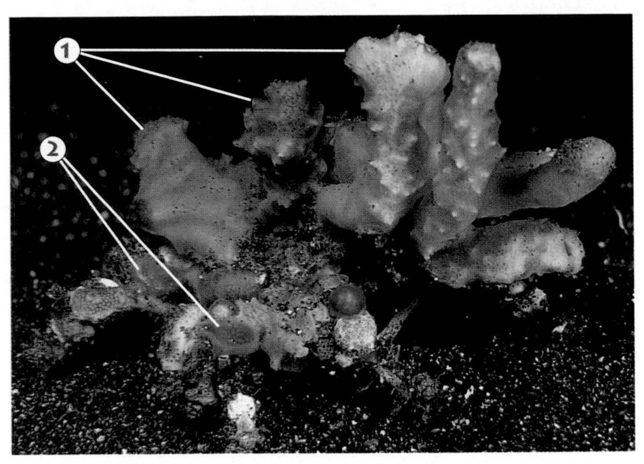

Reef gardener: this hidden decorator crab (*Camposcia retusa*) has transplanted live sponges (**1**), sea squirts (**2**), and other invertebrates to its carapace, making itself virtually undetectable unless it is spotted while moving.

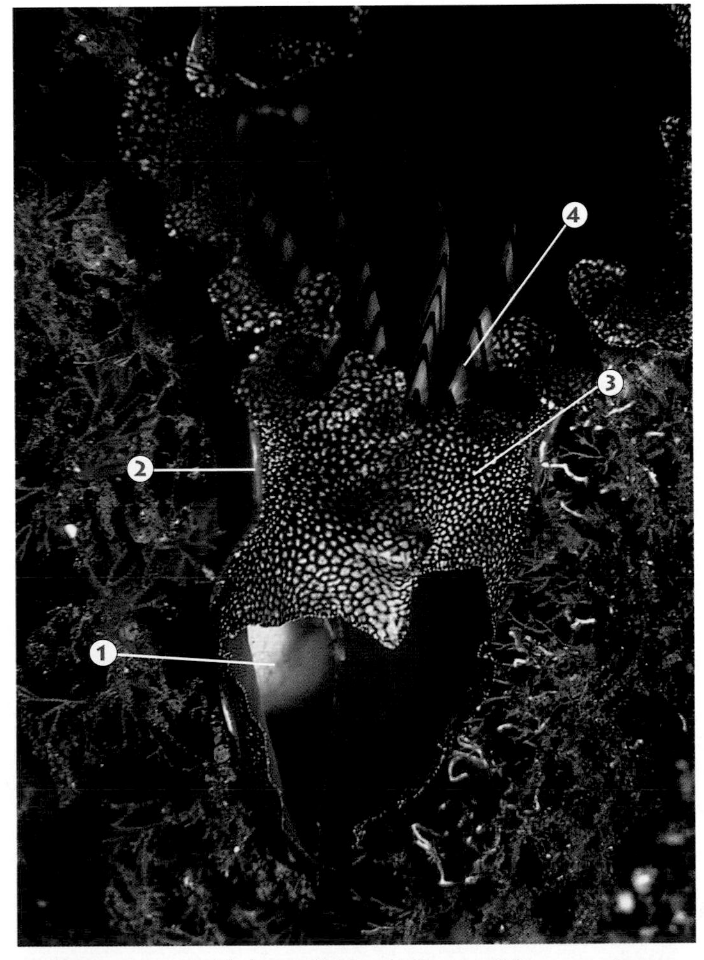

Multiple tactics: this beautiful Rock Oyster (*Chama* sp.) grows in a narrow crevice, protects its soft body (**1**) with a hard shell (**2**), and also displays aposematic (warning) colors on its exposed mantle (**3**), and gills (**4**).

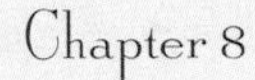

Chapter 8

Reproduction

Procreative Rituals and Strategies

"Nature operates by profusion."
Gabriele Lusser Rico,
Writing the Natural Way (1973)

The small size and secretive behavior of the Green Mandarinfish (Synchiropus splendidus) *make them frustrating photographic subjects, but their brilliant blue, green, and gold colors and mazelike patterns provide ample rewards for those who persevere. During the day they hide within the reef. At dusk they come out for less than an hour to mate and feed. Catching them on film is a challenge, as they shy away from lights, and the low level of natural illumination at dusk makes it difficult to focus the camera. I had been seeking out these elusive fish for some time, and I knew that tonight was the night—I could feel it. I positioned myself and waited for the action to begin.*

Male mandarinfish are larger than females and put on impressive displays to encourage them to mate. I spotted a potential mating pair and concentrated my attention on them. The male followed the female around and flashed his fins at every opportunity. She ignored the amorous show and continued foraging about the reef, but he was not to be distracted from his mission. The male continued to pursue the female until he won her over. At this point they cuddled up, side by side, their naturally protruding lips seemingly puck-

Courting ritual: a pair of Green Mandarinfish (*Synchiropus splendidus*) in full mating colors in a prespawning pose. The male of this species, at left, is larger than the female.

ered as if expecting a kiss, and hovered just above the bottom. They became increasingly aroused as they hovered, and when the time was right, they began their ascent. Fluttering cheek-to-cheek, the pair swam up. The male fanned out his beautiful blue pectoral fins, using them to lift his partner, as if to the stars. At the peak of their ascent, they simultaneously released their sperm and eggs, then descended gently to the bottom as if in a state of rapture. It is sometimes difficult not to draw human parallels when observing reproductive behavior in animals. We are, after all, sexual beings and a little fantasy never hurt anyone. Did I get my sought-after photographs? No, not this time—I was too caught up in the action.

ALL ANIMALS ARE INSTINCTIVELY DRIVEN TO ensure the reproduction of their species. Marine creatures, however, encounter formidable challenges in their pursuit of passing along their genetic heritage. Many of the obstacles they face are unique to their environment, like high reef-population densities and high mortality rates for larvae.

Successful reproduction requires, first and foremost, the task of finding a mate. With the clustering of animals on the coral reef, including members of closely related species living in the same waters, recognizing an appropriate mate is the first prerequisite. Many species synchronize their mating sessions during certain times of day to ensure that breeding males and females find each other in a timely fashion to contribute to the maximum survival of their spawn.

Modes of reproduction on coral reefs are varied and as complex as any found in nature. Although there are not many hard-and-fast rules or generalities about reproduction among reef species, it is true that very few groups are **viviparous**, or **livebearers**, producing young that are fully formed and ready to live independently. Some sharks and rays do give birth to live young, and some corals are able to divide themselves, casting off clones that are immediately recognizable as young, self-sufficient animals.

Most reef fishes and invertebrates are **oviparous**, or egg layers. In turn, this category is divided into three major groups:

Broadcast spawners scatter their eggs freely in the water column. Also called **pelagic spawners,** they release gametes (eggs and sperm) into the water column at times of strong outgoing currents. This sends the eggs seaward where there is less likelihood of predation.

Demersal spawners produce eggs that are attached to the substrate, after which they are sometimes guarded by the parents. Parental care to the degree seen among higher terrestrial animals is the exception, rather than the rule, with marine animals. Some animals will zealously protect their eggs up to the day of hatching, when the larvae drift away and are left to their own devices. Examples of demersal spawners are the anemonefishes, gobies, and triggerfishes.

Brooders provide parental protection for their fertilized eggs as they develop. The mantis shrimp females, for example, carry a sticky mass of eggs between their claws and constantly churn and clean them until they hatch. Many cardinalfishes are **mouthbrooders**, with the male carrying the fertile eggs in his mouth and refusing all food until the young are ready to emerge as free-swimming juveniles.

One reality is the clustering of all manner of predators on, in, and around the reef proper. Eggs and larvae are highly nutritious and eaten with alarming speed and gusto when they become available. The evolutionary responses to this include prodigious fecundity—many marine animals produce hundreds of thousands of gametes each year. Other tactics to overwhelm or elude predators include mass spawning, synchronized spawning, and spending part of their life cycle away from the reef.

Mass spawnings are common, impressive events where large aggregations of animals produce gametes in such huge numbers that at least some are assured of survival. Sessile invertebrates, bound to the reef, cannot move into spawning gatherings but typically synchronize their breeding activities to cast off their gametes, en masse, with other members of their species and, sometimes, spawners from other genera and phyla.

In response to predator pressures, many reef fishes and invertebrates have a two-phase life cycle. They produce **pelagic** (open water) or **planktonic** (drifting with other forms of suspended life) eggs and/or larvae that leave the reef proper temporarily to grow before returning to settle on a reef. Even the young of demersal spawners and brooders may join the off-reef plankton when they have hatched and left their parental care.

Mass spawning: feather stars (Crinoidea) synchronize their spawning for maximum survival of their gametes. Tides, water temperature, phase of the moon, and lighting conditions must be perfect for a synchronized nighttime spawning event.

The concentration of planktivores—smaller fishes and corals, for example—in off-reef surface waters is lower, and predation rates are correspondingly lower. Most pelagic eggs and larvae are adapted to life in the open sea by having flotation devices, body transparency, and a tendency to move toward the light, all of which serve to keep them near the surface and away from predators. When pelagic fish larvae hatch, they have no pigment, no eyes, and no fins. These will develop during the larval stage as they metamorphose into juveniles and begin to look like fishes. Invertebrate development is similar, but may include numerous larval stages with bizarre anatomies and changing feeding habits.

Reproductive timing is highly variable, and marine animals may spawn or give birth seasonally, on a lunar cycle, or even daily. Some tropical species, enjoying a relatively constant environment, spawn throughout the year. Some spawn in spring and fall, while others have an annual reproductive event. Many species coordinate spawning activities with environmentally advantageous periods, such as times of high outgoing currents or abundant food resources for the young larvae. Marine animals are sensitive to slight differences in water temperature, salinity, and the length of the day—changes that frequently signal these favorable conditions. Most reproduction takes place during low-light hours when many planktivores and predators have settled in for the night. Pelagic spawners are more likely to spawn daily at dusk. Demersal spawners are generally on a lunar or semilunar cycle and spawn at dawn. A few species spawn during the day. Some invertebrates and pelagic spawners migrate monthly to special spawning grounds.

Finally, **hermaphrodites**, animals that can produce both eggs and sperm, have their own rules, peculiarities and reproductive rituals—some extraordinarily complex and adaptable to varying conditions. Sexual activity is never far away on the reef, and it's always a temptation to indulge our voyeuristic instincts and take a closer look. ■

Asexual reproduction: a huge, spreading stand of fire coral (*Millepora* sp.) may represent years of cloning, having started as a single sexually produced larva settling on this portion of reef. Such a colony may consist of billions of polyps all traceable to a single ancestor.

GOING WITH THE FLOW

Reproduction among the corals is a fine example of the ability of animals to ensure the survival of their species in the face of many and varied challenges.

Corals are among the most prolific animals on the face of the earth. They have evolved an arsenal of reproductive strategies that all allow them to expand and spread in good times and bad.

Sexual reproduction involves the broadcast spawning of eggs and sperm. Some corals (e.g., *Porites* and *Fungia* spp.) are **gonochoristic**—having separate male and female individuals. **Simultaneous hermaphrodites** (e.g., *Montipora* spp.) are self-fertile and produce **gamete bundles** of eggs and sperm.

In sexual spawning events, huge numbers of gametes are released in the water column to be dispersed off the reef to develop in the plankton. The larval corals may drift for weeks or even several months before settling on a suitable patch of substrate—sometimes hundreds of miles away.

Asexual reproduction in the corals can take several forms. **Fission**, or **budding**, is essentially self-cloning, in which a new polyp is formed when a mature polyp splits in two, drops a small branch, or sprouts a new mouth that develops into a peripheral polyp. **Fragmentation** occurs when stony corals are broken and scattered (as in fierce storms), with new clonal colonies sprouting up where the pieces land. Some corals brood **planulae**, or developed larvae, ready to enter the plankton, while others respond to stress with **polyp bailout**—ejecting free-floating polyps that will settle elsewhere in more favorable conditions. ■

Fission: a tree coral (*Umbellulifera* sp.) reproduces itself asexually by casting off small branchlets or polyp clusters that quickly form new colonies. This coral also engages in sexual reproduction, shedding gametes (visible inside the transparent branches), with other colonies en masse.

New colony: recently shed tree corals anchors themselves promptly to the substrate after being released by their parent. Many such soft corals are prolific in casting off ready-to-grow clones of themselves that drift varying distances before settling and attaching to a hard substrate.

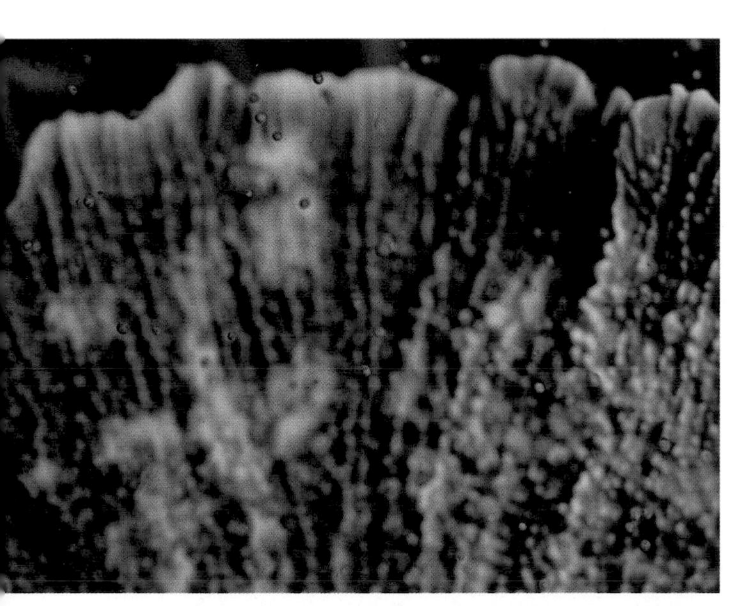

Gamete release: a stony coral (*Montipora* sp.) is caught in the act of releasing gametes into the water column. Although varying scenarios exist, most corals are **hermaphroditic broadcast spawners**, each colony releasing both eggs and sperm, separately or in gamete bundles. Other corals have separate male and female colonies that engage in synchronized spawning events.

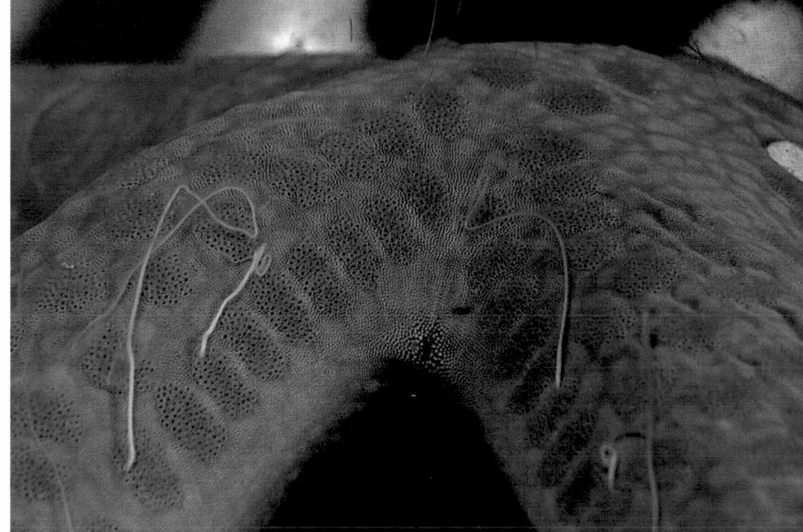

Sexual spawning: a Blue Sea Star (*Linckia laevigata*) rises up on its arm tips and exudes smoky strings of gametes into the water, where fertilization takes place. Most asteroids, or starfish, are **dioecious**—having distinct male and female individuals. Microscopic larvae drift in the plankton until they are ready to settle on a reef. A single female may release up to 2.5 million eggs per year.

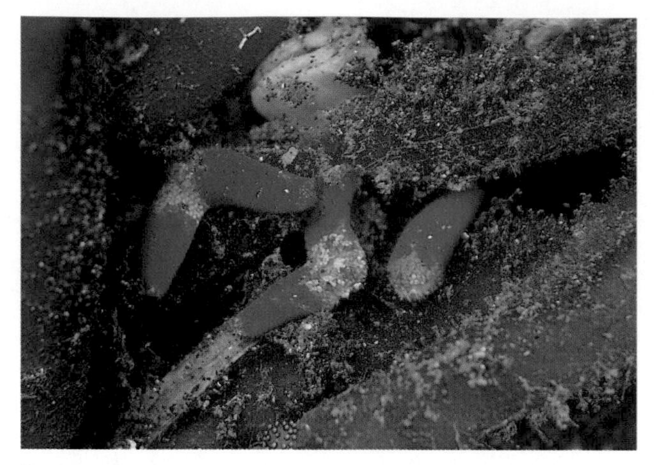

Regeneration: some sea stars, such as the one shown here, reproduce by splitting off arms or larger portions of their body in reproductive events. We have observed mass reproductive regeneration events in Indonesia.

Cloned star: starting with two original arms (the ones on its left side), this new asteroid nears completion of the regeneration process, which can take months. Note the four smaller new arms on the right.

Asexual reproduction: this pulse coral (*Xenia* sp.) creates a new colony by drooping a fleshy extension to the substrate, where it attaches and eventually separates.

Fission: individual stony moon coral polyps (Faviidae) can split, first forming a second mouth and gradually dividing into two adjoining skeletal pockets, or **corallites**.

Fission: solitary sea squirts (*Polycarpa aurata*) generally reproduce by shedding gametes into the water column, but this individual is nearing completion of a self-cloning event, with two nearly identical tunicates emerging from one.

SELF-CLONING

Many invertebrates, most notably echinoderms, have remarkable powers of regeneration—both the ability to regrow lost body parts and to produce clones of themselves. One must be careful not to confuse regeneration and reproduction, although some species do reproduce by casting off and regenerating the missing parts. Regeneration is the regrowth of a body part that has been lost to disease or predation, or one that has been cast off as a defensive measure. **Reproductive regeneration** is the intentional splitting of an organism into two or more parts that will each generate new individuals. In sea star fission, for example, the central disc splits in two and each half regenerates the missing parts. *Linckia* sp. casts off an arm that generates a whole new disc and additional arms. A few species break into several pieces and each generates new body parts as needed.

As explained earlier, corals often produce self-clones by budding, fission, or ejecting fully formed polyps. Fission involves one polyp splitting itself to form two polyps. Generally, the mouth lengthens and eventually closes off in the middle, leaving two separate halves to regenerate the rest of the polyp. Colonies also expand outward by growing buds along the edge or base of the colony. Several species, such as the hard corals *Favia* and *Oculina*, and the soft corals *Dendronephthya* and *Umbellulifera*, are able to cast off a fully developed polyp or group of polyps that go on to settle and develop into a new colony.

Some soft corals have been recorded in the strange act of drooping soft protrusions of tissue to the substrate, where a replica of the original coral springs up. Some sea squirts, as seen above, can also reproduce asexually by budding or by putting out a **stolon**, or rootlike runner of tissue, from which new buds arise.

Many animals employ cloning techniques in addition to other reproductive strategies, often picking the best response to an abundance or shortage of food or changing weather conditions. ■

Mated for life: this breeding pair of Saddleback Anemonefish (*Amphiprion polymnus*) form a lifelong partnership and will lay their eggs in a nest at the base of their host anemone—well protected from potential raiders. The male will guard the nest until the eggs hatch into larvae, which are at the mercy of the currents. Surviving juveniles must find an available host anemone or face sure predation. In a typical large anemone, there may be a number of anemonefish: a dominant matriarch female, her male mate, and one or more small, sexually immature individuals. If the female should die, the male quickly transforms into a functional female, and the highest ranking adolescent moves up to become her mate. Some wits refer to this flexible social unit as "a pair and a spare."

Breeding pair: a small male Giant Frogfish (*Antennarius commerson*) tends his much larger mate until she is ready to rise swiftly to the surface and release a raft of eggs, which he promptly fertilizes. Up to 280,000 floating eggs may be released to develop in the plankton.

Spawning activities: a mated pair of False-eye Puffers (*Canthigaster papua*) in full breeding colors prepare a nesting spot where they will deposit and fertilize a clutch of **demersal** eggs that are placed on the bottom rather than being scattered in the water column.

Simultaneous hermaphrodites: a pair of mating nudibranchs (*Thecacera picta*) copulate, each individual transferring sperm to its partner before separating. Going their own ways, each will produce a fertile mass of eggs. Hermaphroditism—having the organs of both sexes in the same individual—is common among invertebrates.

Spawning aggregation: many marine animals, such as these long-spined red sea urchins (*Astropyga radiata*), gather in large groups for seasonal breeding events. After broadcasting their gametes simultaneously into the water column, they promptly disperse. These gatherings ensure efficient fertilization of the water-borne eggs.

Egg ball: a female Peacock Mantis Shrimp (*Odontodactylus scyallarus*) broods her fertilized eggs that are produced with a sticky adhesive secretion to hold them in a globular mass. The ball is constantly turned and cleaned with her stout raptorial appendages, while she ignores food. Some species produce up to 50,000 eggs at a time. The black dots in the egg mass are the developing eyes of the larval shrimp.

EGGS: PRECIOUS COMMODITIES

Pelagic spawning fishes that abandon their gametes to the tides generally produce more and smaller eggs than the benthic spawners that carefully deposit their spawn on the bottom. While mid-water spawners may easily release tens of thousands of eggs in a single session, a demersal-spawning female may lay a clutch numbering in the hundreds. These eggs, however, are larger and have a significantly higher survival rate, especially with a period of parental care to keep egg-eaters away and to fan and pick away debris and infertile eggs that are attacked by fungus.

Some gastropods lay their eggs en masse, others individually. The egg capsules may contain one or many eggs. Some species brood their eggs with the foot. In a few species, one egg hatches and the larva or juvenile grows by feeding on the remaining undeveloped eggs, a practice known as **oophagy**.

Some marine animals, such as the nudibranchs, produce brightly visible eggs believed to be protected by toxins or bitter tastes. Others produce eggs that are skillfully hidden. Allied and spindle cowries lay their eggs somewhere on their host coral or gorgonian. They strip an area of living coral and deposit the egg mass on the bare branch.

One species of wentletrap is a predator of night-blooming cup corals. The small *Epitonium* kills the polyp and absorbs its yellow pigment. It then deposits tiny yellow crenulated eggs in the skeleton of the coral where they blend in with the living polyps.

Temporary care: a female Titan Triggerfish (*Balistoides viridescens*) tends her egg mass laid in a rocky niche. She will defend the eggs fiercely—even attacking foolhardy divers or snorkelers who enter the nest territory—until they hatch, usually within 24 hours. Once the larvae have emerged, they float off into the plankton.

Blenny eggs on half-shell: a male Jeweled Rockskipper (*Salarias fasciatus*) zealously tends a large mass of fertile eggs laid in an empty clam shell. Such eggs are relished by other reef fishes and must be guarded until the larvae disperse from the nest. Unused mollusk shells are a favorite nesting site of many demersal egg-laying species.

Camouflaged eggs: one of a number of nudibranchs that are virtually indistinguishable from the soft corals on which they live and prey, this *Marionopsis* sp. female lays a red egg ribbon that resembles a small patch of coral. Exposed and vulnerable on the open sand, the eggs may also be protected by toxins or an unpalatable taste.

Fertile eruption: a female box crab (*Calappa* sp.) flicks her abdomen to release thousands of tiny eggs into the currents, where they will hatch and go through a number of stages of development before the survivors settle onto a reef. Gametes, larvae, and juveniles of such invertebrates are vital foods for other reef animals.

Cuttlefish love: a male Broadclub Cuttlefish (*Sepia latimanus*) showing typical noncourtship coloration. Male cuttlefishes must compete aggressively for the opportunity to mate with a female and will use various swimming displays and blazing color patterns—including flashing zebra stripes and iridescence—to win her favor.

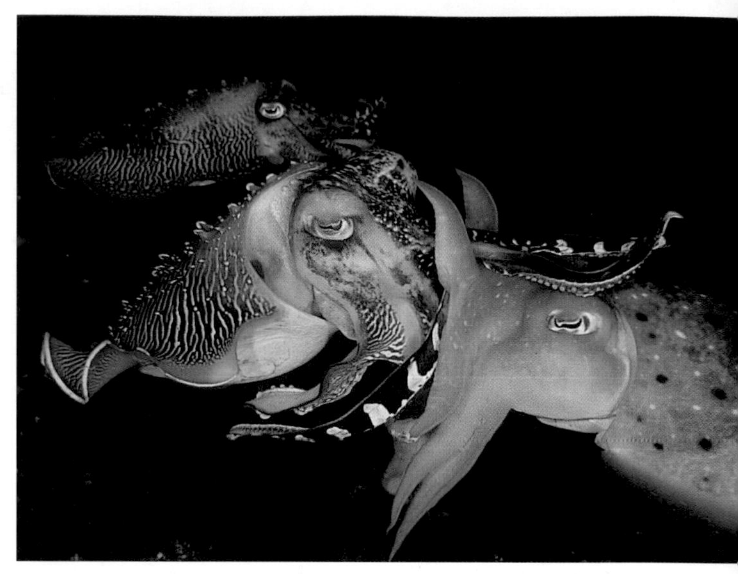

Mating cuttlefish: successful male suitor, left, mates head-on with a female, delivering **spermatophores**, or sperm bundles, into her mantle cavity near her oviducts. Fertilization takes place when the female extrudes her eggs. Note the losing male suitor or "sneaker male" at rear, awaiting any opportunity to mate with the female.

Egg deposition: a female Broadclub Cuttlefish extracts the start of a string of fertilized eggs that she will attach to coral branches or blades of seaweed for safety. Although very intelligent, cuttlefishes have a brief lifespan and the adults typically die shortly after spawning.

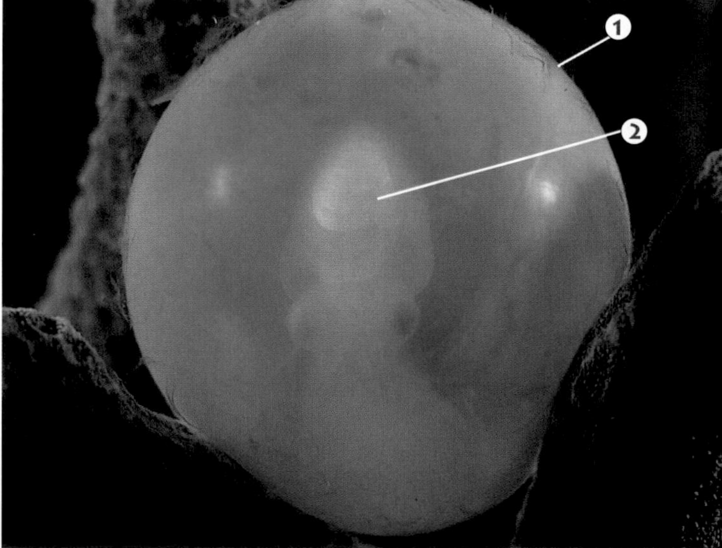

Developing cuttlefish: once laid in a protected site out of the reach of most predators, the eggs swell greatly and develop a tough covering (**1**). Growing in an unusually large egg case with a generous yolk, this *Sepia* sp. embryo (**2**) will hatch as a tiny, perfect cuttlefish (see page 196).

Eggs into hiding: a female cuttlefish carefully deposits her eggs in the relative safety of stony coral branches where they will escape the attention of most passing predators and develop until hatching.

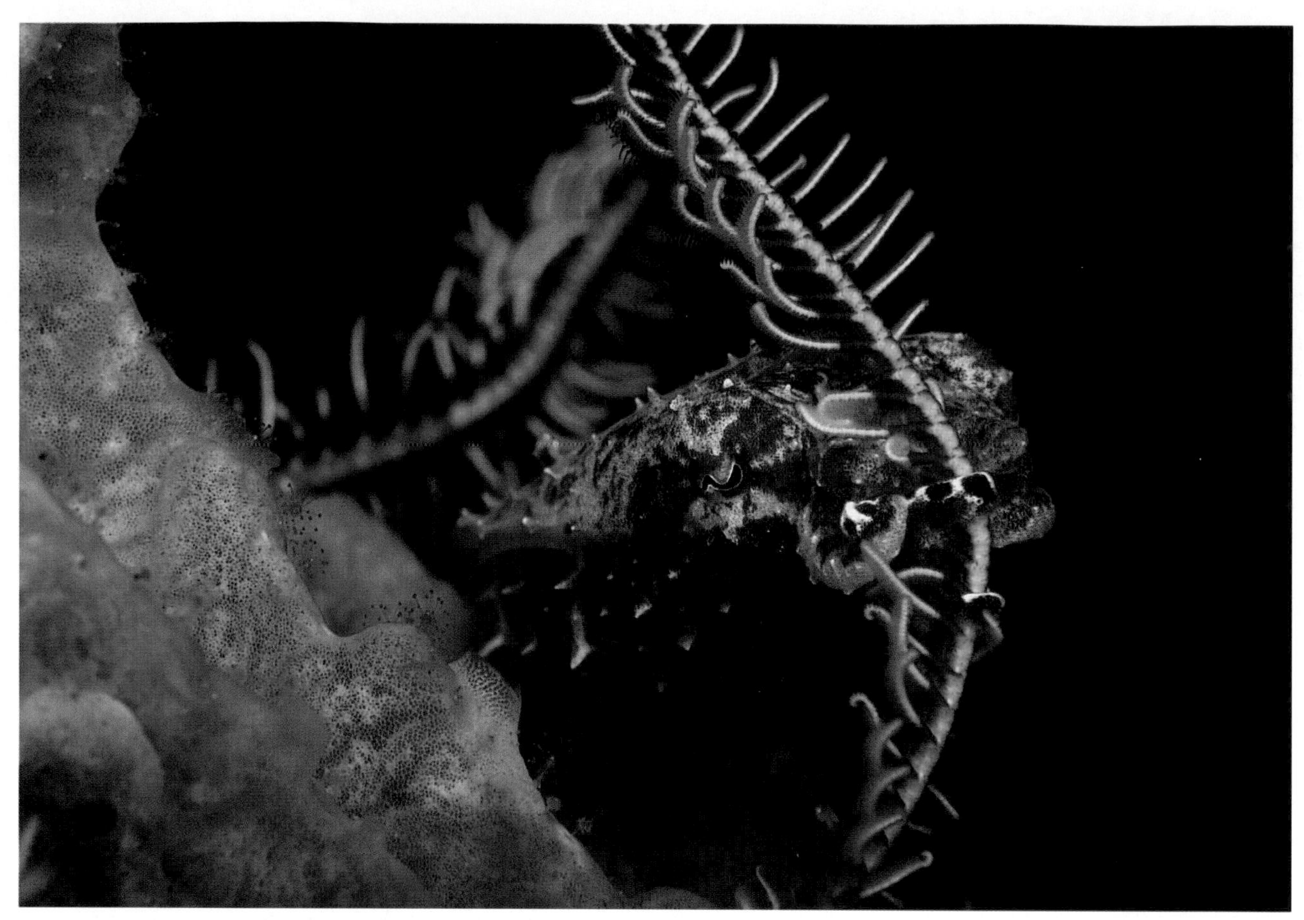

Precarious hold: a tiny, newly settled cuttlefish clutches the arm of a feather star (Crinoidea) for safety during the nighttime hours. Defenseless, such small juveniles are subject to heavy predation until they gain size and strength.

JUVENILES & HERMAPHRODITES

Juvenile fishes and invertebrates often employ extraordinary survival tactics during their vulnerable growth period on the reef. Some display camouflage or warning coloration, others have juvenile patterns that seem to function primarily to let others of their kind know that they are not a competitive threat.

Among the marine angelfishes, for example, many juveniles have no resemblance to their parents. It is believed that this confers some immunity on them from the fierce territoriality of adults (see page 198, bottom left). Only when a fish achieves the strength and size it needs to defend itself does it shift into adult coloration.

Many fishes also delay their emergence as sexual adults in a process known as sequential hermaphroditism. Hermaphrodites can be simultaneous or sequential. **Simultaneous hermaphrodites** can produce both sperm and eggs, but they rarely self-fertilize. Hermaphroditic copulation, with mutual sperm transfer as practiced by nudibranchs, is typical—two animals copulate, with each one transferring sperm and then producing eggs that have been fertilized by the other.

A **sequential hermaphrodite** begins life first as a functional male or female and later changes to the other sex. The major advantages are the flexibility to replace lost producers quickly and the ability to increase the number of breeders as the need arises. Most often the fish begins life as a female in the harem of a dominant male. When the male is removed, for whatever reason, the highest-ranking female moves up to replace him and, in a short time, becomes a fully functioning male. The changeover begins almost immediately and takes a few days to a month to finish. Once complete, the process is usually irreversible.

Although the transformation is a physiological one, it is driven by a change in the social order of the group and serves them well. The gamete producers—the breeding pair—are the biggest and best of the lot, so it follows that their progeny will be of superior quality, thus improving the species stock. ■

Tenuous grip: no bigger than a thimble, this juvenile inflator filefish (*Brachaluteres* sp.) steadies itself in strong currents by simply taking a gorgonian polyp in its mouth and holding on. Larval and juvenile fishes such as this may be carried long distances by ocean currents. Only a small fraction of marine animals survive the lengthy transformation from egg to larva to adult.

Transparency: many larval and postlarval fishes, such as this tiny specimen—likely a wrasse—benefit from having transparent or semi-transparent bodies that allow them to hide with greater ease. Wrasses are sequential hermaphrodites that often develop a drab initial-phase coloration as juveniles, then become females, some of which later transform into colorful terminal-phase males.

Juvenile refuge: young cardinalfish (*Apogon* sp.) seek protective shelter among the stinging tentacles of a Long Tentacle Anemone (*Macrodactyla doreensis*). As with the anemonefishes, these juveniles are able to live in close proximity to the anemone and not trigger its potentially lethal stinging response. Adult cardinalfishes do not typically associate with sea anemones.

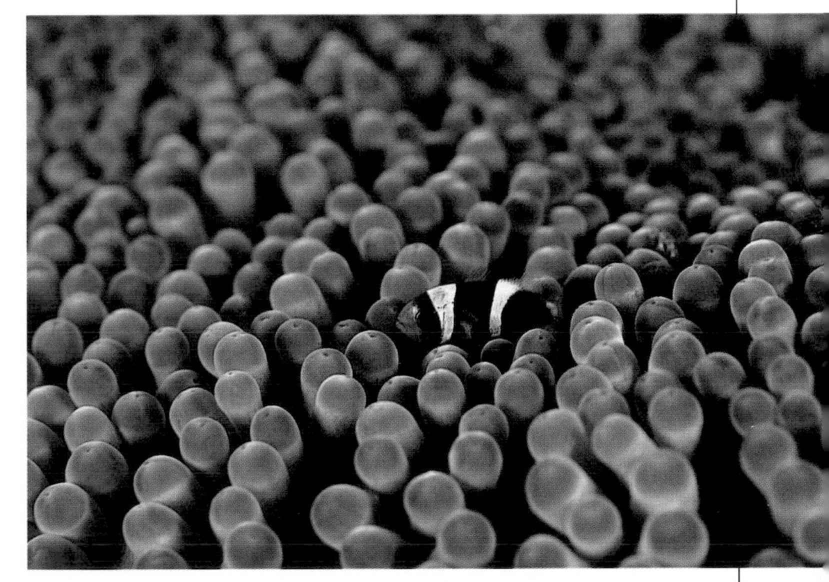

Lifelong safety: this small juvenile Clark's Anemonefish (*Amphiprion clarkii*) nestles into a Bubble Tip Anemone (*Entacmaea quadricolor*) that may serve as its home for life. Once settled into an anemone, these slow-swimming fishes seldom venture very far for fear of falling victim to predators. Metamorphosing juveniles that fail to reach the safety of a host anemone are quickly eaten.

Juvenile colors: a tiny Map Puffer (*Arothron mappa*) displays bright colors that hardly suggest the looks it will achieve as an impressive adult (shown at right) that may reach 65 cm (25 in.) in length. Vivid spots may warn of the toxins that most puffers carry in their skin, gonads, and viscera. Additionally, the yellow-spots-on-black pattern is seen in certain noxious flatworms that predators shun.

Adult colors: a fully mature Map Puffer is a handsome fish with a formidable bite and the ability to inflate itself sufficiently to thwart or deter even large predators. The intricate "map" patterns vary considerably from fish to fish. Although far removed from its juvenile looks, this puffer also carries poisonous tetrodotoxin, and its bold stripes may advertise this fact.

Juvenile pattern: Emperor Angelfish (*Pomacanthus imperator*) bears dazzling markings that gradually fade and reemerge in adult form, shown at right. Marine biologists believe that the distinctive differences between adults and juveniles of such angelfishes confers protection for the youngsters, which would otherwise be mercilessly attacked by adults of their species.

Adult pattern: Emperor Angelfish is considered one of the most beautiful of the many marine angelfishes in its fully mature coloration. They are feisty in defending their territory, especially from other angels and **conspecifics**, or members of their own species. Juvenile Emperors, with their own markings (shown at left), are ranked in a non-threatening social status and thus tolerated.

Before: disruptive coloration of a juvenile Harlequin Sweetlips (*Plectorhinchus chaetodontoides*) bears no resemblance to its future appearance (shown at right). Some biologists speculate that the brash pattern may suggest a toxic flatworm or nudibranch, and the juveniles of this species have an odd, undulating swimming motion that may lend credence to this theory.

After: the adult Harlequin Sweetlips is just one of many medium-sized reef fishes, reaching over 70 cm (27 in.) and 7 kg (15 lbs.), that are important catches for fishermen. It has no apparent visual similarity to the juveniles of its own kind. The evolution of such profound color shifts within species over their lifetimes remains a scientific puzzle.

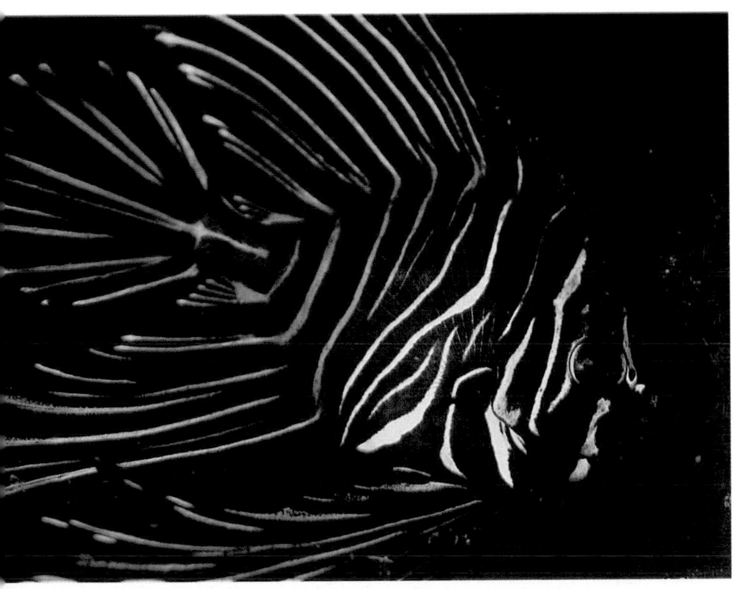

Beauty: the juvenile Zebra Batfish (*Platax batavianus*) is flamboyantly finned and boldly striped, in contrast to its adult form. This coloration may suggest the pattern and flowing fins of a venomous lionfish and thus gain some measure of protection. Alternately, the disruptive color pattern may simply allow the young batfish to hide effectively among seagrasses or mangrove roots.

Beast: an adult Zebra Batfish is also known as the Humphead Spadefish for a prominent hump, or bony brow, that it develops with age. The fish at this stage of its life is no longer vulnerable to many of the smaller reef piscivores that it had to avoid as a juvenile. It is found in deeper waters, away from the inshore shallows where juveniles of the species congregate.

COLOR, SIZE & FECUNDITY

The search for a suitable mate is of utmost importance to marine animals. Coral reef fishes are among the most elaborately ornamented animals on earth. They are often highly patterned and exquisitely colored, but their flashy exterior is not just for show. Colors and patterns can have a number of survival purposes, but they also have a fundamental role in reproduction.

Simply put, unique markings help fishes to identify others of their own species in the maelstrom of life and color on the reef. In groups that have many species with similar markings, such as butterflyfishes, the different species have evolved particular body postures or movements that further identify them to potential mates. Some species recognize potential mates by unique scents, known as pheromones, that they give off just prior to mating time.

Males are generally more colorful than females, although the females tend to be larger because of their egg-carrying capacity. Usually, but not always, it is the males that act to attract the females. The male puts on elaborate displays that include fin flashing, brilliant pattern and color changes, and dancing movements as he tries to entice a female into his territory. Once a female takes notice, the display begins in earnest. The male works frenetically to impress and attract the most desirable female. These high-profile pursuit and display behaviors are risky business in an environment filled with predators.

Fecundity, the potential to produce the most eggs with the best chance of survival, makes females attractive to male suitors. In most marine species, fecundity is directly related to the size of the female, and her size is related to her age. As she grows, her capacity to produce bigger and better eggs increases, and bigger eggs produce bigger larvae that are more likely to survive. Male fishes, unlike many of their human counterparts, have an instinctive tendency to go after the larger, more mature females. ■

Sea of mysteries: Schooling Bannerfish (*Heniochus diphreutes*) and Bluespined Unicornfish (*Naso hexacanthus*) move in opposite directions over a Pacific reef. Despite the harsh odds against their survival from egg to adult, these healthy schools illustrate the ability of thousands of species of reef fishes to reproduce, thrive, and coexist on and around coral reefs.

Epilogue

Disappearing Reefs

Environmental Calamity in Our Own Time

"Only within the moment of time represented by the present century has one species—man—acquired significant power to alter the nature of his world."
Rachel Carson,
Silent Spring (1962)

Reef destruction. Reef decline. Reefs at risk. Doomed reefs. We hear and read the phrases so often but what do they really mean? I am sad to say that we are well acquainted with the deterioration and disappearance of healthy coral reefs. During our many years of working in some of the world's finest diving destinations, we have had the opportunity to dive on reefs that were seldom, if ever, visited by tourists. We have come to realize that pristine reefs are increasingly rare and that the easily accessible ones are often protected for the sport-diving public. Divers, in general, do not see a true picture of the world's reefs because they are under the mistaken impression that all reefs are as healthy as the ones they visit.

The fact is that more than a quarter of the world's coral reefs have been damaged or killed in the last half

Bleaching: an unnaturally white Magnificent Sea Anemone (*Heteractis magnifica*), possibly hundreds of years old, has lost its symbiotic zooxanthellae and color in a tropical warming event caused by El Niño conditions.

Coral harvest: biologically created limestone from the reef is a traditional building material on tropical coasts, but commercial-scale extraction for airport runways, roads, major developments, and mining operations is a serious threat.

century, both from direct local damage and global climatic events. The disappearance or decline of these natural treasures is the culminating result of many human activities that, together, now pose a grave threat to the world's reefs.

When corals die in large numbers, a chain of events is triggered that can lead a vibrant ecosystem supporting huge populations of fishes and invertebrates to deteriorate into a heap of algae-covered rock with a mere fraction of its former life. In bleaching events, when stressed corals suddenly lose or expel their symbiotic zooxanthellae (thus losing color and appearing "bleached"), corals may be killed or seriously damaged. Without living, growing stony corals, the physical structure of the reef begins to collapse. Vital shelter is lost. Animals that feed or live in association with the corals starve or migrate away. Slime bacteria and hair algae often overtake the dead coral skeletons. If the corals do not recover or are not replaced by new colonies dispersed by other reefs, the former beautiful, complex reef settles into a drab mound of fused limestone.

This is not an alarmist fantasy. The statistics of the last few decades are daunting. At the start of this millennium, approximately 11% of the world's coral reefs were classified as dead or so denuded as to have little or no hope of recovery. Another 16% are considered by experts to be degraded, damaged, or under potentially unsustainable stress.

Deforestation, erosion, and coastal development are all sources of deadly sedimentation, allowing silt to cover and smother corals and small invertebrates. (Cloudy water also cuts the penetration of sunlight that corals need to thrive.) Many marine organisms are exquisitely sensitive to chemical pollution, and excessive nutrients from sewage or agricultural runoff can cause algae to proliferate and choke out the corals. Destructive fishing practices can physically destroy reef structure in various ways, or selectively remove certain species and thus disrupt intricate food chains that have existed for millennia.

In Jamaica, for example, live corals covered 52% of the northern coastal reefs in the 1970s; this had dropped to 3% by the early 1990s. The blame has been placed on coastal development, the explosive growth

Spiny plague: hordes of Crown-of-Thorns Starfish (*Acanthaster planci*) (**1**) have stripped the tissue from stony corals (**2**) over large sections of Indo-Pacific reefs in recent years. Hurricanes and typhoons also account for periodic reef damage.

of the tourist industry during this period, and heavy fishing pressure that seriously upset the delicate natural balance of species that healthy reefs achieve.

Burgeoning world population has an easily understood impact on coastal construction, deforestation, water pollution, and overexploitation of the sea. The results are horrific, but the cause and effect are understood, and local remedies can be implemented. In Jamaica, for example, with efforts to protect reef areas, gradual recovery is being achieved, and coral cover—the percentage of the reef substrate occupied by live corals—is back to 10-15% from being virtually gone.

Global warming, on the other hand, raises the spectre of something poorly understood and far from easily corrected. It is the single threat to coral reefs that ecologists fear most. From 1997 through 2000, the Earth suffered the most severe climate changes on record, starting with a major, year-long El Niño tropical warming event followed by a La Niña chill that lasted almost as long. Unnaturally warm water led to worldwide coral bleaching on a scale never before seen. In St. Lucia, coral cover has dropped in recent years from 50% to 25% at 3 m (9.8 ft.) depths, as a result of bleaching and intense hurricanes. In the Indian Ocean, loss of corals to bleaching in the El Niño event of 1997-98 reached 80-90% in the Comoros, Seychelles, and Maldives. Formerly spectacular reefs in Palau, Fiji, and the Solomon Islands have also suffered severe bleaching events in recent years.

While there has been some good news in the recovery of certain areas following a moderation of the water temperatures, some coral biologists now say that the unthinkable is within sight. If global warming trends continue, coral reefs as we know them today could be gone within the lifetime of today's children.

Human impacts on coral reefs and other shallow-water tropical habitats are obvious to anyone willing to look. Over the years we've had the misfortune to witness many assaults on coral reefs and other marine life. On too many reefs there are now very few fishes to be seen. We once lived in a place where the local folks were reduced to living on damselfishes. The bony little damsels were all that were left because the larger species had been fished out long ago.

Occasionally, we have been witness to practices that illustrate how rapacious man can be with life on and around reefs. We once encountered a huge illegal trap net close to shore that had captured whales, dolphins, manta rays, turtles, dugongs, and other hapless animals. It was shut down only after someone went public with a videotape of the crew slaughtering a whale shark. We've made it a habit over the years to empty and disable illegal and abandoned fish traps as we encounter them underwater. While it does one's heart good to set free the trapped fishes, it in no way resolves the problem.

One horrific night, we saw a nesting event of sea turtles—considered threatened throughout the world—turn into a slaughter, as every turtle that came ashore to lay its eggs was killed to be sold as meat. They were never even allowed to lay their eggs. In other countries, we often saw the eggs of sea turtles for sale as delicacies in the local markets.

One of my most cherished memories is seeing a humpback whale and her calf near the reef edge. We were lucky to get close enough for a few underwater photos. Shortly afterward we learned that some local villagers had tried to dynamite the pair. They had been unsuccessful—that time.

Fish bombing is a way of life in large areas of Southeast Asia. Dynamite is the lazy man's way of coming home with a big catch: drop an explosive device into the water and scoop up the dead fish as they rise to the surface. No fuss, no muss. An occasional bomb or two leaves an area of rubble in a stand of established corals. The bombed area will eventually be recolonized by spreading corals and other invertebrates. Repeated use of dynamite, however, reduces the entire area to rubble. Opportunistic species, including algae and soft corals, settle on the rubble and proliferate, eliminating the chance for reef-building stony coral larvae to recolonize the area. Without stony corals, their associated fauna have no place to live. A once-productive area becomes a wasteland.

Bombed reef: unspoiled and bountiful before being reduced to rubble by fish bombs, this reef is typical of many throughout Southeast Asia. Dynamite fishing is illegal but common where enforcement is lax. With coral cover destroyed and the majority of fishes killed or driven away, such areas often become choked with weedlike algal growth, making recovery extremely difficult.

Cyanide has been used for decades to paralyze live fish for food and the aquarium trade, allowing fishermen to capture reef species with relative ease. While sodium or potassium cyanide does not immediately destroy the physical reef structure the way fish bombs do, its effect is no less devastating. Cyanide stuns larger fishes—allowing them to be taken live to very pricey Asian restaurants—but often kills smaller fishes and other reef animals, including corals.

Most countries have enacted laws to protect their reefs from such obvious abuses, but laws are only as good as their enforcement and many countries simply have neither the manpower nor the resources to patrol vast areas of the sea. Large reef areas in the Philippines are in severe decline because of the repeated use of cyanide. After years of trying to curb this practice, a coalition of international groups and reform-minded local governments is finally claiming some success at reducing cyanide fishing. Indigenous fisherfolk are being taught to use nets rather than poisons—and to improve their own livelihoods. The marine aquarium trade is moving toward standards of tracking reef-harvested livestock from reef to end-consumer to ensure that animals are sustainably and properly collected.

Overpopulation is the root cause of most destructive reef practices—too many people, too few resources. So what's the answer? Is there any hope for coral reefs or are they all doomed?

If global warming is caused by human activities, we must act to reverse it. This is the hard part because it calls for an informed public arising against entrenched corporate, energy, and government interests that may well deny global warming until irreversible damage is done.

The need for conservation and population control has never been greater, but there is no one answer that will solve all the problems. It is not enough to tell islanders to stop bombing reefs and slaughtering turtles. We must also provide family planning, and demand sustainably harvested seafoods and certified marine livestock. We must buy tropical woods of known origin to help end rainforest clearcutting. We must stop wasting resources in our personal lives, use less petroleum and fertilizer, and cut the air and water pollution we generate. We must stay informed about the state of the oceans and reefs. We are not ignorant. We know what must be done. The time has come to do it. ■

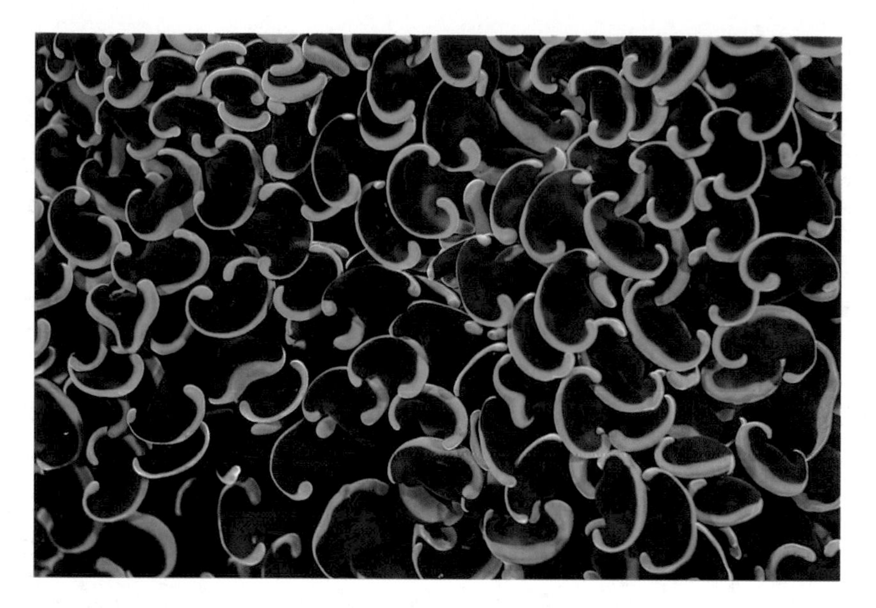

Normally pigmented polyps: this healthy colony of anchor coral (*Euphyllia ancora*) has distinctive polyps heavily populated with zooxanthellae cells—typically golden brown or green—and associated UV-screening pigments. Both the zooxanthellae and the screening pigments can disappear or fade in bleaching events.

Partially bleached polyps: the same species of anchor coral shown above after a bleaching event displays mostly whitened tissue. While these polyps still have the ability to capture plankton and absorb nutrients, they cannot thrive without their symbiotic, energy-generating zooxanthellae. Note golden traces of symbiotic algae cells still remaining in some polyps (**1**).

Bleached colonies: ghostly bleached colonies of anchor coral (*Euphyllia ancora*) demonstrate the response of this species to a period of higher-than-normal temperatures, often associated with windless, becalmed conditions. In this state, the entire patch of coral is malnourished and under extreme stress. If these corals survive, full recovery can take months or years.

Recovery happens: bleaching damage, if the heating event is not severe or prolonged, may be slowly reversed, as in this large, very old colony of mounding *Porites* sp. coral. Dark patches (**1**) indicate that zooxanthellae are starting to return to the stark white coral tissue.

MAJOR REEF THREATS

Human activities considered to be the most significant in causing coral reef decline and disappearance include the following:

Deforestation and erosion can lead to widespread losses of reef systems, which cannot tolerate the influx of silt and sediment that smothers corals. The cutting of tropical rainforests often leads to the destruction of reefs many miles downstream from eroding watersheds that have lost their tree cover.

Coastal development is a multi-pronged problem, often starting with the clearing of mangroves and shoreline construction that triggers erosion into formerly clean waters. Development brings pollution in the form of wastewater, desalination effluents, sewage, agricultural fertilizers, and industrial chemicals.

Overexploitation includes the wholesale taking of corals and coral rock for road, runway, and building construction, as well as the making of cement. Unsustainable harvest of live marine resources by fishing and trapping is a significant problem in areas with no effective fisheries management.

Destructive fishing practices, such as fish bombing and use of poisons, typically potassium or sodium cyanide, kill corals and destroy natural fish balances.

Marine pollution from ports and ship traffic has serious detrimental impacts on coral reefs.

Global warming may be the final blow for widespread areas of coral reefs, which cannot adapt quickly enough to changing water temperatures. ■

Treacherous trash: a Cockatoo Waspfish (*Ablabys taenianotus*) swallows a bristleworm along with strands of plastic. Such man-made materials can cause fatal blockages in the digestive systems of marine animals.

GLOBAL WASTE

Long ago, humans who lived or travelled in coastal areas learned that if they tossed something into the sea, it went away and usually did not come back. For shoreline dwellers and ships at sea, this seemed to be the perfect trash-disposal system. It worked in the past because discards were materials that Nature could dissolve, erode, and hence recycle. Since the advent of plastics, disposable diapers, and the like, this is no longer true. The refuse becomes permanent.

We watched one lovely reef disappear as the local population covered it with old refrigerators and washers and then packed garbage into the remaining spaces as they tried to expand the size of their island.

We have seen many other disturbing sights underwater: fishing nets irretrievably snagged on the reef and continuing to trap fishes and crabs that struggle to free themselves before dying; mollusks ensnared in nets that prevent them from fully exposing their mantles; small sharks fighting to free themselves from gill nets; fishes with hooks broken off in their mouths; corals broken by carelessly tossed boat anchors; garbage; household discards and old clothes washed up on otherwise pristine beaches; drifting plastic bags; big black oil globules on white sand beaches; cans, batteries, bottles, toys. The sad truth is that persistent trash and raw sewage pour into the tropical seas daily, altering the nutrient balance and compromising the health of many reefs. Marine life can adapt to make use of some trash items such as glass bottles, but how can it adapt to spilled oil or golf course pesticides?

This wanton waste problem is global, affecting coral reefs and shallow-water environments throughout much of the world, not just the Tropics. ■

Flotsam and jetsam: a string of garbage washed up on the beach of an uninhabited Pacific island. Even remote tropical islands are now strewn with a worldwide plague of litter from garbage dumped into the sea. Among the culprits are freighters, cruise ships, and coastal communities that use the ocean as a convenient trash can.

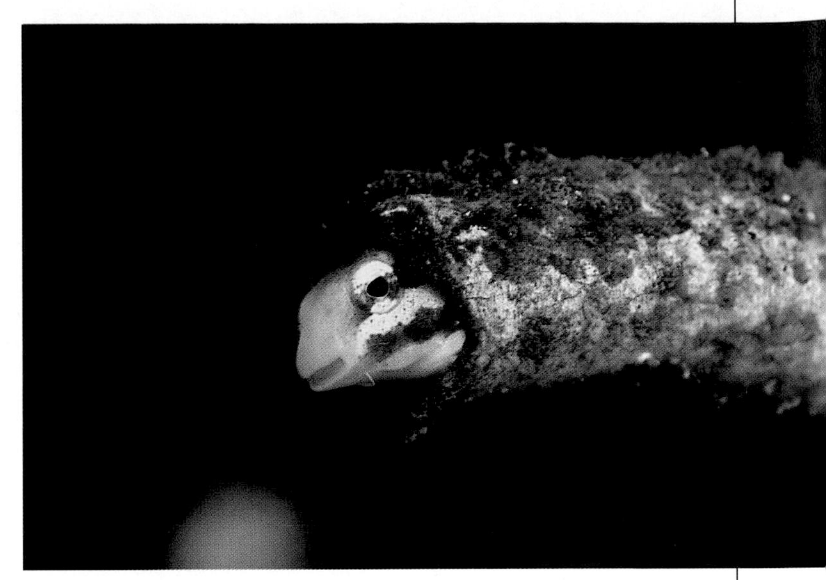

Home from the range: a Striped Fang Blenny (*Meiacanthus grammistes*) makes its home in the spout of a rusting teapot that has landed miles from the nearest kitchen. Some human discards can be adapted as habitat by reef dwellers that move in for shelter or colonize every available surface.

Synthetic shield: a small octopus (*Octopus* sp.) hides in a coconut shell and not very successfully attempts to cover itself with a plastic plate from a tape cassette, part of a growing plague of human discards dumped at sea by ships and growing populations in coastal locations.

Ghost nets: this young shark illustrates the all-too-common result of becoming entangled in an abandoned fishing net. Hundreds of tons of such "ghost nets" are drifting loose in the world's oceans and draping the substrate in still-lethal material.

GLOSSARY

AGGRESSIVE MIMICRY: a type of camouflage in which an aggressive animal mimics a benign animal or environmental feature to take advantage of an unwary animal for self gain.

ALGAL RIDGE: a ridge of coralline algae that absorbs some of the wave shock along the outer edge of some reefs.

ALLELOPATHY: the emission of chemicals by one species that affects the behavior or physiology of another species in the same environment.

ANAEROBIC: without oxygen.

APHOTIC: without light.

APOSEMATIC COLORATION: advertising one's unpalatability or toxicity by displaying bright or bold colors and patterns.

ASEXUAL REPRODUCTION: reproduction that takes place without the production of gametes, e.g., fission, budding.

ATOLL: a ring or partial ring of coral topped by a sandy island that surrounds an interior lagoon.

AUTOTOMY: the intentional casting off of an appendage, often at predefined points; self-amputation.

AUTOTROPH: organisms that use sunlight and absorbed inorganic molecules to produce organic materials via photosynthesis.

BANK REEF: coral formations running parallel to a coastline, separated from the shore by a channel that is relatively narrow and shallow (classic example: Florida Reef Tract); compare to **BARRIER REEF.**

BARRIER REEF: massive coral formation running parallel to a coastline, separated from the shore by a relatively deep and wide expanse of ocean (classic example: Great Barrier Reef); compare to **BANK REEF.**

BATESIAN MIMICRY: a type of imitation in which a harmless animal mimics a harmful animal, often by duplicating its aposematic coloration, to gain protection from predators.

BENTHIC: relating to the seafloor; bottom-dwelling.

BINOMIAL NOMENCLATURE: a two-part system of scientific naming, using the genus and species.

BIOME: environmentally distinct ecological community.

Coleman's Shrimp (*Periclimenes colemani*) with parasitic isopods in its gills rests atop a toxic sea urchin (*Asthenosoma* sp.).

BIVALVE: a mollusk with two shells (valves), e.g., clams, oysters, mussels.

BLEACHING: when stressed corals suddenly lose or expel their symbiotic zooxanthellae, thus appearing to be whitened or "bleached."

BROADCAST SPAWNERS: species that scatter their eggs and sperm freely in the water column.

BROODERS: species that provide parental protection for fertilized eggs as they develop.

BUDDING: the asexual creation of an animal by the formation of a small growth attached to the adult; the bud may stay attached or may be released to form a new colony.

CAMOUFLAGE: the act of blending in with one's environment.

CARAPACE: the hard outer covering, or shell, of crustaceans.

CARNIVORE: animal-eating.

CARRION: dead or decaying tissue.

CARTILAGINOUS: having a body shaped by connective tissue instead of bone, e.g., sharks, rays.

CAUDAL: in reference to the tail.

CEPHALOPODS: mollusks in which the foot has been formed into arms or tentacles surrounding the head, e.g., octopuses, squids, cuttlefishes.

CERATA: projections along the dorsal surface of many nudibranchs.

CHAETAE: the bristles of polychaete worms.

CHELIPED: a clawed appendage of crustaceans.

CHEMORECEPTOR: a specialized cell that responds to chemical signals in the environment.

CHROMATOPHORE: a specialized pigmented skin cell that can be expanded or contracted to display or hide its color.

CILIA: movable, hairlike projections.

CIRRI: small, often curly, appendages on many animals.

CNIDARIANS: radially symmetrical invertebrates with stinging structures called nematocysts.

CNIDOCYTES: specialized cells of cnidarians that contain stinging structures called nematocysts.

CNIDOSAC: tip of a projection from the body of a cnidarian-eating nudibranch that contains nematocysts acquired from its prey.

COLOR MORPH: distinctive color variant of the same species.

COMMENSALISM: a symbiotic relationship in which one party (the symbiont) benefits without harming the other party (the host).

CONSPECIFIC: within the same species.

COPEPOD: minute, free-living crustacean.

COUNTERSHADING: coloration that is dark on the top side and light on the under side.

CREPUSCULAR: active at dawn and/or dusk.

CROSS-FERTILIZATION: a type of reproduction in which hermaphrodites copulate, injecting each other with sperm and fertilizing each other's eggs.

CRYPSIS: the combination of camouflage and cryptic (concealment) behavior.

CRYPTOFAUNA: small, hidden animal life.

CUTICLE: the external, nonliving layer of skin.

DECOY MIMICRY: see **AGGRESSIVE MIMICRY.**

DEMERSAL: bottom-dwelling.

DEMERSAL SPAWNERS: species that attach their eggs to the substrate.

DETRITIVORE: an animal that feeds on animal and/or plant waste material (detritus) that has settled to the bottom.

DETRITUS: dead animal and plant material, along with the bacteria that decompose it.

DIFFUSION: the tendency of particles to move randomly to occupy an available space.

DINOFLAGELLATES: one-celled organisms in the Kingdom Protista that possess characteristics of both plants and animals.

DIOECIOUS: having distinct male and female individuals.

DIURNAL: active during the daylight hours.

DOM: dissolved organic matter; organic material dissolved in seawater.

DORSAL: the back or upper surface of an animal.

ELASMOBRANCH: member of the Subclass Elasmobranchii; a shark or ray.

ELECTROCYTES: specialized cells, evolved from muscle cells, that concentrate and channel electrical signals.

ELECTROLOCATION: locating an object, traversing a course, or communicating by the aid of low-frequency electrical stimuli.

EMERSION: the act of freshwater or marine animals being exposed to air.

EPIBENTHIC: describing an organism that lives in the water column but stays very close to the bottom.

EPIBIONT: a living organism (nonparasitic) that settles on the surface of another living organism.

EPITOKE: a pelagic reproductive individual produced by a sessile or benthic individual; characteristic of many polychaete worms.

EQUILIBRIUM: a state of balance.

ESCA: in frogfishes, the fleshy tip of the first dorsal spine (the illicium); often looks like a small fish or worm and is used in luring behavior.

ESTUARY: coastal inlet where freshwater streams or rivers meet the ocean.

EUTROPHICATION: accelerated level of plant or algal growth in a body of water resulting in the depletion of dissolved oxygen.

EVERT: to turn inside out.

EVISCERATE: to expel part of the internal organs.

EXTENSIBLE: capable of being extended.

EYEBAR: a dark bar extending over the eye of a fish that tends to camouflage the presence of the eye.

FAUNA: animal life.

FECUNDITY: a measure of the reproductive potential of a female.

FISSION: asexual reproduction of an organism by splitting into two or more pieces, each of which will generate a new organism.

FLORA: plant life.

FRINGING REEF: a mass of coral growing in shallow waters adjacent to the shore of a continental landmass or island.

GAMETES: cells specialized for fertilization, i.e., sperm and eggs.

GASTROPOD: one- or no-shelled mollusk, e.g., snails, sea slugs, nudibranchs.

GILL RAKERS: projections along the gills, elongated for straining organisms from water in filter-feeding species of fishes.

GONAD: an organ that produces sperm or eggs.

HAREM: a group of females under the reproductive control of one male.

HERBIVORE: plant-eating animal.

HERMAPHRODITE: an animal with both male and female reproductive organs.

HETEROTROPH: an animal that feeds primarily or exclusively on plant material.

HOST: the organism with which a symbiont has a relationship.

ILLICIUM: modified first dorsal spine of a frogfish that is elongate and usually adorned with a fleshy tip (the esca); used in luring behavior.

INFERIOR: underneath; on the bottom of.

INITIAL-PHASE MALE: subordinate male that often resembles a female in coloration; some go on to become supermales.

INSTINCT: behavior that is stereotypical and inherited; it is neither learned nor changed by experience.

INTERSPECIFIC: between different species.

INTERSTITIAL FAUNA: animal life occurring in the water between grains of sand.

INVERTEBRATE: an animal without a backbone.

IRIDOCYTE: a type of color cell in the skin that produces white, silvery, or iridescent effects.

IRIDOPHORE: a chromatophore containing structures that reflect light, making it appear silvery or iridescent.

LARVA: independent, preadult form of an animal, usually occurs just after hatching from an egg; it is markedly different from the adult in appearance.

LATERAL: referring to the side or sides.

LATERAL LINE: sensory system in fishes in which pressure changes are detected by cells in the skin along both sides of the body.

MACROALGAE: large, multicellular algae; can be red, green, or brown.

MARINE SNOW: drifting bits of organic matter suspended in the water column.

MEDUSA: free-swimming form of cnidarians, e.g., jellyfishes.

MEIOFAUNA: tiny animals that live in the interstitial spaces on the seafloor.

MESENTERIAL FILAMENTS: long tubes or tentacles that extend from the gut of cnidarians and are used in feeding.

METAMORPHOSE: to change or transform from a larva to an adult.

MIMICRY: the act of imitating another species in appearance and/or behavior in order to hide one's true appearance.

MÜLLERIAN MIMICRY: a form of mimicry in which different species, both unpalatable or dangerous, mimic each other.

MUTUALISM: a commensal relationship in which both parties benefit.

NEMATOCYSTS: the stinging structures inside the cnidocyte cells of cnidarians.

NEUROTOXIN: a toxin that works on the nervous system of an organism.

NICHE: the physical range and faunal hierarchy in which an organism can live.

NOCTURNAL: active at night.

OBLIGATE SYMBIONT: a symbiont that cannot live without its host and is never found without it.

OCELLUS: an eyespot (plural, **OCELLI**).

OMNIVORE: an organism that eats a variety of materials, both animal and vegetable.

OOPHAGY: feeding on eggs.

OPISTHOBRANCHS: gastropod mollusks with greatly reduced or no shell that have not undergone torsion.

OSCULA: the excurrent openings of sponges.

OSTIA: the small incurrent openings of sponges.

OVIPARITY: a type of reproduction in which eggs are released and develop outside the body of either parent.

PARASITE: a symbiotic organism that obtains nourishment at the expense of its host.

PARASITISM: a symbiotic relationship in which one party (the parasite) benefits, while the other party (the host) is harmed.

PATCH REEF: a reef that develops independently of a main framework; often scattered near shore or in shallow waters.

PELAGIC: living in the water column.

PHARYNGEAL: relating to the throat; behind the mouth cavity.

PHEROMONES: scents released by animals that enable them to communicate with others; many pheromones are sexual attractants.

PHOTIC: having light.

PHOTOSYNTHESIS: the process by which plants, with the aid of sunlight, turn carbon dioxide, water, and other molecules into carbohydrates.

PHYTOPLANKTON: microscopic algae in the water column.

PIGMENT: the color-containing substance of chromatophores.

PLANKTIVORE: an organism that feeds on plankton.

PLANKTON: microscopic plants and animals in the water column.

PLANULAE: cnidarian larvae; usually elongate and flattened.

POLYCLAD: having more than one gut; referring to flatworms in the Order Polycladida.

POLYP BAILOUT: coral ejection of free-floating polyps.

POM: particulate organic matter; component of marine snow.

PROBOSCIS: an extensible mouth tube.

PROSTOMIUM: flat proboscis of an echiuran, or spoon worm.

PROTANDRY: a form of sequential hermaphroditism in which the organism is first a male and later changes to a female.

PROTOGYNY: a form of sequential hermaphroditism in which the organism is first a female and later changes to a male.

PROTRUSIBLE: capable of being extended.

RADIOLE: each of the pinnate tentacles on the head of a polychaete worm.

RADULA: a membranous structure in mollusks containing transverse rows of teeth.

RAPTORIAL APPENDAGES: the chelipeds (claws) of stomatopods (mantis shrimps) used for grasping prey.

REEF CREST: the upper part of the reef slope just over the edge from the reef flat.

REEF SLOPE: the outer part of the reef that angles down from the reef flat to the seafloor.

REGENERATION: the act of regrowing.

REPRODUCTIVE REGENERATION: an asexual form of reproduction in which an organism casts off part of itself and each part then generates a whole individual, e.g., some sea stars cast off an arm that goes on to regenerate a new central disc and the missing arms.

RHINOPHORE: nudibranch tentacles located near the head that are considered to be olfactory organs.

ROSTRUM: an anterior middorsal projection of crustacean carapaces; snout.

SACOGLOSSAN: herbivorous opisthobranch.

SCAVENGER: an organism that feeds on carrion.

SCHOOL: social group of fishes, usually but not always of the same species but of approximately equal social rank and size (compare to **SHOAL**).

SCLERACTINIAN: stony coral.

SEMINAL RECEPTACLE: the chamber in female animals that receives and stores sperm.

SENESCENCE: old age.

SEQUENTIAL HERMAPHRODITISM: a form of hermaphroditism in which an animal is first either a functional male or female and later in life changes to become a functional member of the opposite sex.

SESSILE: attached; not free-swimming.

SETAE: bristles or stiff hairs of echinoderms.

SETTLING OUT: the process by which pelagic juvenile animals go from being neutrally buoyant to negatively buoyant, resulting in their drift down to the seafloor.

SHADOW: to swim closely alongside another fish so as to be unnoticed.

SHOAL: an aggregation of fishes, all of the same species but not necessarily of the same size and social rank, that swim in an organized fashion (compare to **SCHOOL**).

SIMULTANEOUS HERMAPHRODITISM: a form of reproduction in which each animal can produce both sperm and eggs but it rarely self-fertilizes.

SIPHON: a tubular extension that allows water to flow in and out of an organism.

SNEAKER: term applied to some male organisms that sneak in to mate with a female while her defending male is occupied elsewhere.

SOCIALIZATION: behavioral modifications that result from interactions with other organisms.

SOM: suspended organic matter, also called pseudoplankton; component of marine snow.

SPECIOSE: having many species.

SPERMATOPHORE: packet of sperm.

SPICULES: calcareous (calcium carbonate-based) or siliceous (silica-based) skeletal parts of sponges, corals, and other invertebrates; often needlelike.

SPIROCYSTS: adhesive cells along the tentacles of anemones that aid in capturing prey.

STOLON: a horizontal stem or extension of the body along which new buds or individuals form.

SUBORDINATE MALE: a male fish that is subordinate to a supermale.

SUBSTRATE: foundation upon which organisms grow.

SUBTERMINAL: situated near to but not at the end; in fishes, often refers to the location of the mouth as being just under the snout.

SUPERMALE: a terminal-phase male that usually dominates a harem of females and displays filamentous dorsal rays and bright colors.

SWEEPER TENTACLES: elongate tentacles of some corals that contain cnidocytes and can be used in aggressive or defensive interactions.

SYMBIONT: an organism living in a relationship with another organism of a different species.

SYMBIOSIS: two different organisms living together in a relationship.

TACTILE: pertaining to touch.

TAXA: groupings of life within the Linnaean taxonomic system of classification (singular, **TAXON**).

TENDRIL: a grasping or attaching appendage.

TENTACULAR: having tentacles.

TERMINAL-PHASE MALE: a supermale.

TEST: the shell of an invertebrate.

TUBICOLOUS: tube-dwelling.

VENTRAL: the underneath or bottom side.

VERTEBRATE: animal with a backbone.

VESICLE: a membranous and usually fluid-filled pouch in a plant or animal.

VIVIPARITY: a type of reproduction in which the young develop inside the parent; young are born fully formed.

ZOOPLANKTON: microscopic animals in the water column.

ZOOXANTHELLAE: dinoflagellate photosynthetic algae that are symbionts and live within the tissues of many invertebrates, especially corals.

BIBLIOGRAPHY

Abramson, C.I. 1994. *A Primer of Invertebrate Learning: The Behavioral Perspective.* American Psychological Assn., Washington, D.C.

Allen, G. 1997. *Marine Fishes of South-east Asia,* 3rd ed. Western Australian Museum, Perth, Australia.

Allen, G. 1979. *Butterfly and Angelfishes of the World, Vol. 2.* Mergus Publishers, Melle, Germany.

Allen, G.R., and R. Steene. 1994. *Indo-Pacific Coral Reef Field Guide.* Tropical Reef Research, Singapore.

Anderson, R.C., and J.A. Mather. 1996. Escape responses of *Euprymna scolopes* Berry, 1911 (Cephalopoda: Sepiolidae). *Journal of Molluscan Studies,* 62 (4): 543-545, Nov.

Anderson, R.C. *Living Reefs of the Maldives.* Novelty Press, Malé, Maldives.

Bakus, G.J., et al. 1994. *Coral Reef Ecosystems.* A.A. Balkema Publishers, Rotterdam.

Becker, K., and M. Wahl. 1996. Behaviour patterns as natural antifouling mechanisms of tropical marine crabs. *Journal of Experimental Marine Biology and Ecology,* 203 (2):245-258, 15 Oct.

Bergquist, P.R. 1978. *Sponges.* University of California, Berkeley.

Berrill, N.J. 1950. *The Tunicata, With an Account of the British Species.* Ray Society, London.

Bliss, D.E. 1982. *Shrimps, Lobsters and Crabs.* Columbia University Press, N.Y.

Bliss, D.E., and L.H. Mantel, eds. 1985. *The Biology of Crustacea, Vol. 9.* Academic Press, Orlando, FL.

Bond, C.E. 1996. *Biology of Fishes,* 2nd ed. Saunders College Publishing, Ft. Worth, TX.

Borneman, E.H. 2001. *Aquarium Corals: Selection, Husbandry, and Natural History.* T.F.H. Publications Inc., Neptune City, NJ and Microcosm Ltd., Charlotte, VT.

Breakthroughs: Fiber-optic Sponges. 1997. *Discover* 18 (3):23, Mar.

Brusca, R.C. 1980. *Common Intertidal Invertebrates of the Gulf of California.* University of Arizona, Tucson.

Bryant, D., et al. 1998. *Reefs at Risk: A Map-Based Indicator of Threats to the World's Coral Reefs.* World Resources Institute, Washington, D.C.

Castro, P., and M.E. Huber. 1992. *Marine Biology.* Wm. C. Brown Publishers, Dubuque, IA.

Colin, P.L. 1988. *Marine Invertebrates and Plants of the Living Reef.* T.F.H. Publications, Inc., Neptune City, NJ.

Colin, P.L., and C. Arneson. 1995. *Tropical Pacific Invertebrates.* Coral Reef Press, Beverly Hills, CA.

Coleman, N. 1997. *The Dive Sites of the Great Barrier Reef and the Coral Sea.* Passport Books, Lincolnwood, IL.

Dahan, M., and Y. Benayahu. 1997. Reproduction of *Dendronephthya hemprichii* (Cnidaria:Octocorallia): year-round spawning in an azooxanthellate soft coral. *Marine Biology,* 129 (4):573-579.

Darwin, C. 1984. *The Structure & Distribution of Coral Reefs.* The University of Arizona Press, Tucson, AZ.

David, B., et al., eds. 1994. *Echinoderms through Time.* A.A. Balkema Publishers, Rotterdam.

Day, J.H. 1967. *A Monograph on the Polychaeta of Southern Africa.* Trustees of the British Museum, London.

Debelius, H. 1996. *Nudibranchs and Sea Snails, Indo-Pacific Field Guide.* IKAN-Unterwasserarchiv, Frankfurt, Germany.

Dionisio-Sese, M.L., et al. 1997. UV-absorbing substances in the tunic of a colonial ascidian protect its symbiont, *Prochloron* sp., from damage by UV-B radiation. *Marine Biology,* 128 (3):455-461.

Encarta Encyclopedia 1997. CD-ROM. Microsoft, Redmond, WA.

Encyclopaedia Britannica 1998. CD-ROM. Encyclopaedia Britannica, Inc., Chicago.

Fauchald, K. 1977. *The Polychaete Worms, Definitions and Keys to the Orders, Families and Genera.* Natural History Museum of L.A. County, Los Angeles.

Forsythe, J.W., and R. T. Hanlon. 1997. Foraging and associated behavior by *Octopus cyanea* Gray, 1849 on a coral atoll, French Polynesia. *Journal of Experimental Marine Biology and Ecology,* 209 (1-2):15-31, Feb.

George, D., and J. George. 1979. *Marine Life: An Illustrated Encyclopedia of Invertebrates in the Sea.* John Wiley and Sons, N.Y.

Glausiusz, J. 1997. The Importance of Being Infected. *Discover* 18:30, May.

Gochfeld, D.J., and G.S. Aeby. 1997. Control of populations of the coral feeding nudibranch *Phestilla sibogae* by fish and crustacean predators. *Marine Biology,* 130 (1):63-69.

Godin, J.G.J. 1997. *Behavioural Ecology of Teleost Fishes.* Oxford University Press, Oxford.

Gosliner, T. 1987. *Nudibranchs of Southern Africa.* Sea Challengers, Monterey, CA.

Gosliner, T.M., et al. 1996. *Coral Reef Animals of the Indo-Pacific.* Sea Challengers, Monterey, CA.

Griffith, J.K. 1997. Occurrence of aggressive mechanisms during interactions between soft corals (Octo-corallia:Alcyoniidae) and other corals on the Great Barrier Reef, Australia. *Marine and Freshwater Research,* 48 (2):129-135.

Guilcher, A. 1988. *Coral Reef Geomorphology.* John Wiley and Sons, Ltd. Chichester, U.K.

Gulko, D. 1998. *Hawaiian Coral Reef Ecology.* Mutual Publishing, Honolulu, HI.

Hanlon, R.T., and J.B. Messenger. 1996. *Cephalopod Behaviour.* Cambridge University Press, Cambridge.

Harbison, G.R. 1985. On the Classification and Evolution of the Ctenophora. In: S.C. Morris, et al., eds. *The Origins and Relationships of Lower Invertebrates, The Systematics Association Special Volume 28.* Clarendon Press, Oxford.

Helmuth, B.S.T., et al. 1997. Interplay of host morphology and symbiont microhabitat in coral aggregations. *Marine Biology,* 130 (1):1-10.

Hendler, G., et al. 1995. *Sea Stars, Sea Urchins and Allies: Echinoderms of Florida and the Caribbean.* Smithsonian Institution Press, Washington, D.C.

Kaplan, E.H. 1982. *Peterson Field Guides, Coral Reefs.* Houghton Mifflin, Boston.

Kramarsky-Winter, E. 1997. Popping polyps. *Discover* 18 (9):15, Sept.

Kuiter, R.H. 1992. *Tropical Reef-Fishes of the Western Pacific, Indonesia and Adjacent Waters.* Penerbit PT Gramedia Pustaka Utama, Jakarta, Indonesia.

Kuiter, R.H., and H. Debelius. 1994. *Southeast Asia, Tropical Fish Guide.* IKAN-Unterwasserarchiv, Frankfurt, Germany.

Lagler, K.F., et al. 1977. *Ichthyology,* 2nd ed. John Wiley and Sons Inc., New York.

Levine, J.S., and J.L. Rotman. 1985. *Undersea Life.* Stewart, Tabori and Chang, New York.

Lieske, E., and R. Myers. 1994. *Reef Fishes of the World.* Periplus Editions (HK) Ltd., Hong Kong.

Lieske, E., and R. Myers. 1999. *Coral Reef Fishes: Caribbean, Indian Ocean, and Pacific Ocean, Including the Red Sea.* Princeton University Press, Princeton, NJ.

McKinsey, K. 1998. Dances of Worms. *Scientific American,* pp. 28, May.

McMillan, B., and J.A Musick. 1997. *Oceans, Life in the Deep.* Metro Books, New York.

Michael, S.W. 2001. *Reef Fishes, Vol. 1.* T.F.H. Publications Inc., Neptune City, NJ and Microcosm Ltd., Charlotte, VT.

Milius, S. 1998. Hermaphrodites duel for manhood. *Science News,* 153:101, Feb. 14.

Mlot, C. 1997. Clockwork sex of coral reef algae. *Science News,* 151:134, Mar. 1.

Morris, D. 1990. *Animal Watching.* Crown Publishers, New York.

Moyle, P.B. 1993. *Fish, An Enthusiast's Guide.* University of California Press, Berkeley.

Moyle, P.B., and J.J. Cech, Jr. 1996. *Fishes, An Introduction to Ichthyology,* 3rd ed. Prentice-Hall, Upper Saddle River, NJ.

Myers, R.F. 1989. *Micronesian Reef Fishes.* Coral Graphics, Guam.

Nielsen, T.M. 1982. *The Marine Biology Coloring Book.* HarperCollins, New York.

Olson, R.R. 1984. The Life History and Larval Ecology of the Ascidian–Algal Symbiosis, *Didemnum molle.* (Thesis 8419456, Harvard University, Ph. D.)

Paxton, J.R., and W.N. Eschmeyer, eds. 1995. *Encyclopedia of Fishes.* Academic Press, San Diego, CA.

Pitcher, T.J., ed. 1993. *Behaviour of Teleost Fishes,* 2nd ed. Chapman and Hall, London.

Pörtner, H.O., et al. 1994. *Physiology of Cephalopod Molluscs, Lifestyle and Performance Adaptations.* Gordon and Breach Science Publishers, Amsterdam.

Randall, J.E. 1968. *Caribbean Reef Fishes.* T.F.H. Publications, Inc., Neptune City, NJ.

Randall, J.E., et al. 1990. *Fishes of the Great Barrier Reef and Coral Sea.* University of Hawaii Press, Honolulu.

Reaka-Kudla, M.L., et al., eds. 1997. *Biodiversity 11, Understanding and Protecting Our Biological Resources.* Joseph Henry Press, Washington, D. C.

Ruppert, E.E., and R.D. Barnes. 1991. *Invertebrate Zoology,* 6th ed. Saunders College Publishing, Ft. Worth, TX.

Sale, P.F., ed. 1991. *The Ecology of Fishes on Coral Reefs.* Academic Press, San Diego, CA.

Scalera-Liaci, L., and C. Canicatti, eds. 1991. *Echinoderm Research.* A.A. Balkema Publishers, Brookfield, VT.

Sheppard, C.R.C. 1983. *A Natural History of the Coral Reef.* Blandford Press, Hong Kong.

Smith, M.M., and P.C. Heemstra. 1986. *Smith's Sea Fishes.* Southern Book Publishers, Johannesburg, South Africa.

Sprung, J. 2001. *Invertebrates: A Quick Reference Guide.* Ricordea Publishing, Miami, FL.

Stancyk, S.E. 1979. *Reproductive Ecology of Marine Invertebrates.* University of South Carolina, Columbia.

Steene, R.C. 1977. *Butterfly and Angelfishes of the World, Volume 1, Australia.* Mergus Publishers, Melle, Germany.

Tomascik, T., et al. 1997. *The Ecology of the Indonesian Seas, Parts 1 and 2.* Periplus Editions (HK) Ltd., Singapore.

Tompa, A.S., et al., eds. 1984. *The Mollusca, Volume 7, Reproduction* (K.M. Wilbur, ed. in chief). Academic Press Inc., Orlando, FL.

Trueman, E.R., and M.R. Clarke, eds. 1988. *The Mollusca, Volume 11, Form and Function* (K.M. Wilbur, ed. in chief). Academic Press Inc., San Diego, CA.

Valentine, J.W., and E.M. Moores. 1982. Plate Tectonics and the History of Life in the Oceans. In A.T. Newberry, ed., *Scientific American, Life in the Sea.* W. H. Freeman and Co., San Francisco. pp. 19-28.

Vernberg F.J., and W.B. Vernberg. 1983. *The Biology of Crustacea, Volume 7, Behavior and Ecology* (D. E. Bliss, ed.). Academic Press, New York.

Vernberg F.J., and W.B. Vernberg, eds. 1983. *The Biology of Crustacea, Volume 8, Environmental Adaptations* (D.E. Bliss, ed. in chief). Academic Press, New York.

Veron, J.E.N. 1986. *Corals of Australia and the Indo-Pacific.* University of Hawaii Press, Honolulu.

Veron, J.E.N. 1995. *Corals in Space & Time: The Biogeography and Evolution of the Scleractinia.* Comstock/Cornell, Cornell University Press, New York.

Wainwright, P.C., and S.M. Reilly, eds. 1994. *Ecological Morphology, Integrative Organismal Biology.* University of Chicago, Chicago.

Walls, J.G., ed. 1982. *Encyclopedia of Marine Invertebrates.* T.F.H. Publications, Inc., Neptune City, NJ.

Warner, R.R. 1986. The Environmental Correlates of Female Infidelity in a Coral Reef Fish. In T. Uyeno, et al. eds., *Indo-Pacific Fish Biology–Proceedings of 2nd International Conference on Indo-Pacific Fishes.* Ichthyological Society of Japan, Tokyo. pp. 803-810.

Wells, F.E., and C. W. Bryce. 1993. *Sea Slugs of Western Australia.* Western Australian Museum, Perth, Australia.

Wells, M.J. 1978. *Octopus, Physiology and Behaviour of an Advanced Invertebrate.* Chapman and Hall Ltd., London.

Wells, M.J. 1962. *Brain and Behaviour in Cephalopods.* Heinemann Educational Books, London.

Wilson, B., et al. 1993. *Australian Marine Shells, Prosobranch Gastropods, Part One.* Odyssey Publishing, Kallaroo, Australia.

Wilson, E.O. 1980/1995. *Sociobiology. The Abridged Edition.* Belknap Press of Harvard University Press, Cambridge, MA.

Wilson, R., and J.Q. Wilson. 1985. *Watching Fishes.* Harper and Row, New York.

Wood, E.M. 1983. *Corals of the World.* T.F.H. Publications Inc., Neptune City, NJ.

Wu, C. 1997. Crab crackers. *Science News,* 151:122-123, Feb. 22.

Zann, L.P. 1980. *Living Together in the Sea.* T.F.H. Publications Inc., Neptune City, NJ.

INTERNET SOURCES

Borneman, Eric. "Getting Up-To-Date on Zooxanthellae." Online. Internet. 12 Dec 2001. Available FTP: aquarium.net.

"Biodiversity: Relative Numbers of Described Species in Major Taxa." Online. World Resources Institute. Internet. 11 Dec. 2001. Available FTP: wri.org.

"Oceans and Coasts Global Marine Strategy." Online. World Resources Institute. Internet. 11 Dec 2001. Available FTP: wri.org.

"Coral Reef Biodiversity: Teaming with Biodiversity." Online. Coral Reef Alliance. Internet. 12 Dec 2001. Available FTP: coralreefalliance.org/aboutcoralreefs/reefbiodiversity.

"How Many Species Are There?" Online. World Resources Institute. Internet. 11 Dec 2001. Available FTP: wri.org.

"New Atlas Maps World's Fast Disappearing Coral Reefs." Online. United Nations Environment Programme. Internet. 12 Dec. 2001. Available FTP: coralreefalliance.org/ press/unepatlas.

"Biodiversity and Conservation: How Many Species Are There?" Online. Offwell Woodland & Wildlife Trust. Internet. 11 Dec 2001. Available FTP: offwell.free-online.co.uk/biodvy.

"Blue Planet Challenge". Online. BBC Homepage. 10 Dec 2001. Available FTP: bbc.co.uk/blueplanet.

"Some Toxic Algae Cause Thousands of Seafood Poisonings Worldwide." SC Sea Grant Consortium—Toxic Algae Newsletter. Online. Internet. 10 Dec 2001. Available FTP: scseagrant.org/library.

"Dazzling Diversity: Biological Diversity in the Marine Environment." Online. World Resources Institute. Internet. 11 Dec 2001. Available FTP: wri.org.

"Tropical Forest Species Richness." Online. World Resources Institute. Internet. 11 Dec. 2001. Available FTP: wri.org.

INDEX

NUMBERS IN BOLDFACE INDICATE PHOTOGRAPHS

ABOUT THE AUTHORS

IN THE EARLY 1980S, Larry and Denise Tackett left established careers in chemical engineering and banking for part-time jobs collecting marine specimens for a university in search of new anti-cancer drug leads. They chose not to work on a full-time basis in order to spend more time doing what they loved diving, observing, and photographing nature.

Throughout the course of their university work, the Tacketts lived in remote areas with primitive conditions in the Indo-Pacific. Diving for work and pleasure, they logged thousands of dives in Truk Lagoon, Palau, Papua New Guinea, Maldives, Seychelles, Comoros, Mauritius, Malaysia, Singapore, and Indonesia.

Several years ago they left the university to pursue photography and nature writing full time. Larry and Denise are contributing editors to *Skin Diver* magazine. Denise co-authored several editions of the travel book *Insight Pocket Guide: Maldives*.

The Tacketts are represented by stock photo agencies in the U.S., Canada, and United Kingdom, and their photographs and articles have been widely published in books and magazines worldwide. They also teach photography and lead seminars and workshops internationally. The Tacketts were featured in a segment of *Ushuaïa Nature*, for French television, and were the subject of a National Geographic *Sea Stories* episode featuring their work on pygmy seahorses and Lembeh Strait.

Larry and Denise spend several months a year traveling and gathering new material. When in the U.S., they reside in wild, wonderful West Virginia.

Their Web site is: http://tackettproductions.com.